BASIC  PROBABILITY

What Every Math Student Should Know

Second Edition

BASIC  PROBABILITY

What Every Math Student Should Know

Second Edition

Henk Tijms

 **World Scientific**

NEW JERSEY · LONDON · SINGAPORE · BEIJING · SHANGHAI · HONG KONG · TAIPEI · CHENNAI · TOKYO

Published by

World Scientific Publishing Co. Pte. Ltd.

5 Toh Tuck Link, Singapore 596224

USA office: 27 Warren Street, Suite 401-402, Hackensack, NJ 07601

UK office: 57 Shelton Street, Covent Garden, London WC2H 9HE

Library of Congress Cataloging-in-Publication Data

Names: Tijms, H. C., author.

Title: Basic probability : what every math student should know / Henk Tijms.

Description: Second edition. | New Jersey : World Scientific, [2021] | Includes index.

Identifiers: LCCN 2021012561 | ISBN 9789811237492 (hardcover) |
    ISBN 9789811238512 (paperback) | ISBN 9789811237508 (ebook)

Subjects: LCSH: Probabilities--Textbooks.

Classification: LCC QA273.2 .T54 2021 | DDC 519.2--dc23

LC record available at https://lccn.loc.gov/2021012561

British Library Cataloguing-in-Publication Data

A catalogue record for this book is available from the British Library.

For any available supplementary material, please visit
https://www.worldscientific.com/worldscibooks/10.1142/12295#t=suppl

Printed in Singapore

Preface

Probability is a fascinating branch of mathematics and indispensable for statistical literacy. In modern society it is necessary to be able to analyze probability claims in the media and to make informed judgments and decisions. This book provides a basic introduction to probabilistic thinking and reasoning under uncertainty, and illustrates it with many instructive and interesting examples.

All the essential aspects of basic probability are covered in this book. It grew out of a probability book I wrote for Dutch high school students and it is written for students learning about probability for the first time. The book is aimed specifically at statistics and data science students who need an accessible and solid introduction to the basics of probability. Probability is the bedrock for statistics and data science.

Key features of the book are conditional and Bayesian probability, real-world applications of the Poisson distribution, and the interplay between probability and computer simulation. Computer simulation is not only a computational tool for solving probability problems, but it is also a useful didactic tool for helping students to gain a better understanding of probabilistic ideas. The emphasis of the book is on discrete probability, but continuous probability distributions are also treated. Scattered throughout the text, many remarks are made on the history of probability.

Because probability is a difficult subject for the beginner, the approach followed in this book is to develop probabilistic intuition before getting into details. Whereas many probability problems are easy to understand, the solutions can require clever and creative thinking. The best way to learn probability is by doing a lot of problems. Many instructive problems together with problem-solving strategies are given. Answers to all problems and worked-out solutions to selected problems are provided to increase students' confidence in problem solving and to stimulate students' active learning.

Preface to the Second Edition

In the second edition many didactic improvements have been made. Also, many extra motivational examples and problems have been added. The material on Bayesian probability has been expanded by two famous court cases. The chapter on real-life applications of the Poisson distribution has been expanded by two new sections. Sections on generating functions and the bivariate normal distribution have been added. The roles of the law of large numbers and the central limit theorem in computer simulation have been discussed in more depth. The most significant change is the addition of an introductory chapter on Markov chains, including absorbing Markov chains and Markov chain Monte Carlo simulation. This appealing material, which is popular with students, can be addressed directly after the notion of conditional probability has been discussed.

Finally, thanks to all those who have made valuable comments on the first edition. In particular, I would like to thank Karl Sigman and Alan Vraspir.

Contents

Chapter 1

Combinatorics and Calculus for Probability

This chapter presents a number of results from combinatorics and calculus, in preparation for the subsequent chapters. Section 1.1 introduces you to the concepts of factorials and binomial coefficients. In Section 1.2 the exponential function and the natural logarithm will be discussed.

1.1 Factorials and binomial coefficients

Many probability problems require counting techniques. In particular, these techniques are extremely useful for computing probabilities in a chance experiment in which all possible outcomes are equally likely. In such experiments, one needs effective methods to count the number of outcomes in any specific event. In counting problems, it is important to know whether the order in which the elements are counted is relevant or not. Factorials and binomial coefficients will be discussed and illustrated.

In the discussion below, the *fundamental principle of counting* is frequently used: if there are a ways to do one activity and b ways to do another activity, then there are $a \times b$ ways of doing both. As an example, suppose that you go to a restaurant to get some breakfast. The menu says pancakes, waffles, or fried eggs, while for a drink you can choose between juice, coffee, tea, and hot chocolate. Then the total number of different choices of food and drink is $3 \times 4 = 12$. As another example, how many different license plates are possible when the license plate displays a nonzero digit, followed by three letters, followed by three digits? The answer is $9 \times 26 \times 26 \times 26 \times 10 \times 10 \times 10 = 158\,184\,000$ license plates.

Factorials and permutations

How many different ways can you order a number of different objects such as letters or numbers? For example, what is the number of different ways that the three letters A, B, and C can be ordered? By writing out all the possibilities ABC, ACB, BAC, BCA, CAB, and CBA, you can see that the total number is 6. This brute-force method of writing down all the possibilities and counting them is naturally not practical when the number of possibilities gets large, as is the case for the number of possible orderings of the 26 letters of the alphabet. You can also determine that the three letters A, B, and C can be ordered in 6 different ways by reasoning as follows. For the first position, there are 3 available letters to choose from, for the second position there are 2 letters left over to choose from, and only one letter for the third position. Therefore the total number of possibilities is $3 \times 2 \times 1 = 6$. The general rule should now be evident. Suppose that you have n distinguishable objects. How many ordered arrangements of these objects are possible? Any ordered sequence of the objects is called a *permutation*. Reasoning in the same way as above gives that there are n ways for choosing the first object, leaving $n - 1$ choices for the second object, etc. Therefore the total number of ways to order n *distinguishable* objects is equal to the product $n \times (n-1) \times \cdots \times 2 \times 1$. This product is denoted by $n!$ and is called 'n factorial'. Thus, for any positive integer n,

$$n! = 1 \times 2 \times \cdots \times (n-1) \times n.$$

A useful convention is

$$0! = 1,$$

which simplifies the presentation of several formulas to be given below. Note that $n! = n \times (n-1)!$ and so $n!$ grows very quickly as n gets larger. For example, $5! = 120$, $10! = 3\,628\,800$ and $15! = 1\,307\,674\,368\,000$. Summarizing, for any positive integer n,

the total number of ordered sequences (permutations) of n distinguishable objects is $n!$.

Armed with this knowledge, you are asked to argue that in the lottery 6/45 the six lotto numbers, which are drawn one by one from 1 to 45, will appear in either an ascending or descending order with a probability of $\frac{1}{360}$.

An interesting question is how many different words, whether or not existing, can be constructed from a given number of letters where some letters appear multiple times. For example, how many different words can be constructed from five letters A, two letters B, two letters R, one letter C, and one letter D? To answer this question, imagine that the five letters A are labeled as A_1 to A_5, the two letters B as B_1 and B_2, and the two letters R as R_1 and R_2. Then you have 11 different letters and the number of ways to order those letters is 11!. The five letters A_1 to A_5, the two letters B_1 and B_2, and the two letters R_1 and R_2 can among themselves be ordered in $5! \times 2! \times 2!$ ways. Each of these orderings gives the same word if you replace A_1 to A_5 by A, B_1 and B_2 by B, and R_1 and R_2 by R. Thus the total number of different words that can be formed from the original 11 letters is

$$\frac{11!}{5! \times 2! \times 2!} = 83\,160.$$

Thus, if you thoroughly mix the eleven letters and then put them in a row, the probability of getting the word ABRACADABRA is equal to $\frac{1}{83\,160}$.

Binomial coefficients and combinations

How many different juries of three persons can be formed from five persons A, B, C, D, and E? By direct enumeration you see that the answer is 10: $\{A, B, C\}$, $\{A, B, D\}$, $\{A, B, E\}$, $\{A, C, D\}$, $\{A, C, E\}$, $\{A, D, E\}$, $\{B, C, D\}$, $\{B, C, E\}$, $\{B, D, E\}$, and $\{C, D, E\}$. In this problem, the order in which the jury members are chosen is not relevant. The answer 10 juries could also have been obtained by a basic principle of counting. First, count how many juries of three persons are possible when attention is paid to the order. Then determine how often each group of three persons has been counted. Thus the reasoning is as follows. There are 5 ways to select the first jury member, 4 ways to then select the next member, and 3 ways to select the final member. This would give $5 \times 4 \times 3$ ways of forming the jury when the order in which the members are chosen would be relevant. However, this order makes no difference. For example, for the jury consisting of the persons A, B and C, it is not relevant which of the $3!$ ordered sequences ABC, ACB, BAC, BCA, CAB, and CBA has led to the jury. Hence the total number of ways a jury of 3 persons can be formed from a group of 5 persons is equal to $\frac{5 \times 4 \times 3}{3!}$. This expression can be rewritten as

$$\frac{5 \times 4 \times 3 \times 2 \times 1}{3! \times 2!} = \frac{5!}{3! \times 2!}.$$

In general, you can calculate that the total number of possible ways to choose a jury of k persons out of a group of n persons is equal to

$$\frac{n \times (n-1) \times \cdots \times (n-k+1)}{k!}$$
$$= \frac{n \times (n-1) \times \cdots \times (n-k+1) \times (n-k) \times \cdots \times 1}{k! \times (n-k)!} = \frac{n!}{k! \times (n-k)!}.$$

This leads to the definition

$$\boxed{\binom{n}{k} = \frac{n!}{k! \times (n-k)!}}$$

for non-negative integers n and k with $k \le n$. The quantity $\binom{n}{k}$ (pronounce: n choose k) has the interpretation:

> $\binom{n}{k}$ **is the total number of ways to choose k different objects out of n distinguishable objects, paying no attention to their order.**

In other words, $\binom{n}{k}$ is the total number of combinations of k different objects out of n and is referred to as the *binomial coefficient*. The key difference between permutations and combinations is *order*. Combinations are *unordered* selections, permutations are *ordered* arrangements.

The binomial coefficients play a key role in the so-called *urn model*. This model has many applications in probability. Suppose that an urn contains R red and W white balls. What is the probability of getting exactly r red balls when blindly grasping n balls from the urn? To answer this question, it is helpful to imagine that the balls are made distinguishable by giving each of them a different label.[1] The total number of possible combinations of n different balls is $\binom{R+W}{n}$. Under these combinations there are $\binom{R}{r} \times \binom{W}{n-r}$ combinations with exactly r red balls (and thus $n - r$ white balls). Thus, if you blindly grasp n balls from the urn, then

$$\text{the probability of getting exactly } r \text{ red balls} = \frac{\binom{R}{r} \times \binom{W}{n-r}}{\binom{R+W}{n}}$$

with the convention that $\binom{a}{b} = 0$ for $b > a$. These probabilities represent the so-called *hypergeometric distribution*. Probability problems that can be translated into the urn model appear in many disguises. A nice illustration is the lottery 6/45. In each drawing of the lottery, six different numbers are chosen from the numbers $1, 2, \ldots, 45$. Suppose you have filled in one ticket with six distinct numbers. Then the probability of matching exactly r of the drawn six numbers is

$$\frac{\binom{6}{r} \times \binom{39}{6-r}}{\binom{45}{6}} \quad \text{for } r = 0, 1, \ldots, 6,$$

as you can see by identifying the six drawn numbers with 6 red balls and the other 39 numbers with 39 white balls. In particular, the probability of matching all six drawn numbers (the jackpot) equals 1 to 8 145 060.

In mathematics there are many identities in which binomial coefficients appear. The following recursive relation is known as Pascal's triangle:[2]

$$\binom{n}{k} = \binom{n-1}{k-1} + \binom{n-1}{k} \quad \text{for } 1 \leq k \leq n.$$

---

[1] Labeling objects to distinguish them from each other can be very helpful when solving a combinatorial probability problem.

[2] Pascal was far from the first to study this triangle. The Persian mathematician Al-Karaji had produced something very similar as early as the 10th century, and the triangle is called Yang Hui's triangle in China after the 13th century Chinese mathematician Yang Hui, and Tartaglia's triangle in Italy after the 16th century Italian mathematician Niccolò Tartaglia.

You can algebraically prove this. A more elegant proof is by interpreting the same 'thing' in two different ways. This is called a word-proof. Think of a group of n persons from which a committee of k persons must be chosen. The k persons can be chosen in $\binom{n}{k}$ ways. However, you can also count as follows. Take a particular person, say John. The number of possible committees containing John is given by $\binom{n-1}{k-1}$ and the number of possible committees not containing John is given by $\binom{n-1}{k}$, which verifies the identity.

Try yourselves the following test questions:

• How many distinct license plates with three letters followed by three digits are possible? How many if the letters and numbers must be different? (answer: $17\,576\,000$ and $11\,232\,000$).

• What is the number of ways to arrange 5 letters A and 3 letters B in a row? (answer: 56)

• Five football players A, B, C, D and E are designated to take a penalty kick after the end of a football match. In how many orders can they shoot if A must shoot immediately after C? How many if A must shoot after C? (answer: 24 and 60)

• What is the total number of distinguishable permutations of the eleven letters in the word Mississippi? (answer: $34\,650$).

• John and Pete are among 10 players who are to be divided into two teams A and B, each consisting of five players. How many formations of the two teams are possible so that John and Pete belong to a same team? (answer: 112)

• Suppose that from 10 children, five are to be chosen and lined up. How many different lines are possible? (answer: $30\,240$).

• Give word proofs of $\binom{n}{n-k} = \binom{n}{k}$ and $\sum_{k=0}^{n} \binom{n}{k}\binom{n}{n-k} = \binom{2n}{n}$.

• How many ways are there to distribute eight identical chocolate bars between five children so that each child gets at least one chocolate bar? (answer: 35)[3]

---

[3]The number of combinations of *non-negative* integers $x_1, \ldots, x_n$ satisfying $x_1 + \cdots + x_n = r$ is $\binom{n+r-1}{r}$. This result is stated without proof.

1.2 Basic results from calculus

The history of the number e begins with the discovery of logarithms by John Napier in 1614. At this time in history, international trade was experiencing a period of strong growth, and, as a result, there was much attention given to the concept of compound interest. At that time, it was already noticed that $(1 + \frac{1}{n})^n$ tends to a certain limit if n is allowed to increase without bound:

$$\lim_{n \to \infty} \left(1 + \frac{1}{n}\right)^n = e \text{ with } e = 2.7182818\ldots.$$

The famous mathematical constant e is called the Euler number. This constant is named after Leonhard Euler (1707–1783) who is considered as the most productive mathematician in history.

The *exponential function* is defined by e^x, where the variable x runs through the real numbers. This is one of the most important functions in mathematics. A fundamental property of e^x is that this function has itself as derivative:

$$\frac{de^x}{dx} = e^x \quad \text{for all } x.$$

Intermezzo: Let's sketch a derivation. Consider a function $a(x)$ of the specific form $a(x) = a^x$ for some constant $a > 0$. Then, for each $h > 0$,

$$\frac{a(x+h) - a(x)}{h} = \frac{a^{x+h} - a^x}{h} = a(x)\frac{a^h - 1}{h}.$$

Take for granted that $c = \lim_{h \to 0}(a^h - 1)/h$ exists. Thus $a'(x) = ca(x)$. When is the constant $c = 1$? The answer is if $a = e$. To see this, note that the condition $\lim_{h \to 0}(a^h - 1)/h = 1$ is the same as $a = \lim_{h \to 0}(1 + h)^{1/h} = \lim_{n \to \infty}(1 + \frac{1}{n})^n = e$. This shows that $a'(x) = a(x)$ if $a = e$. A more general result is that $f(x) = e^x$ is the only function satisfying the differential equation $f'(x) = f(x)$ with the boundary condition $f(0) = 1$.

How to calculate the function e^x? The generally valid relation

$$\lim_{n \to \infty} \left(1 + \frac{x}{n}\right)^n = e^x \quad \text{for all } x$$

is not useful for that purpose. The calculation of e^x is based on the power series

$$e^x = 1 + x + \frac{x^2}{2!} + \frac{x^3}{3!} + \cdots \qquad \text{for all } x.$$

The proof of this power series expansion requires Taylor's theorem from calculus. The fact that e^x has itself as derivative is crucial in the proof. Note that term-by-term differentiation of the series $1 + x + \frac{x^2}{2!} + \cdots$ leads to the same series, in agreement with the fact that e^x has itself as derivative.

The series expansion of e^x leads to $e^x \approx 1 + x$ for x close to 0. This is one of the most useful approximation formulas in mathematics! In probability theory the formula is often used as

$$e^{-x} \approx 1 - x \qquad \text{for } x \text{ close to 0.}$$

A nice illustration of the usefulness of this formula is provided by the birthday problem. What is the probability that two or more people share a birthday in a randomly formed group of m people (no twins)? To simplify the analysis, it is assumed that the year has 365 days (February 29 is excluded) and that each of these days is equally likely as birthday. Number the people as 1 to m and let the sequence $(v_1, v_2, \ldots, v_m)$ denote their birthdays. The total number of possible sequences is $365 \times 365 \times \cdots \times 365 = 365^m$, while the number of sequences in which each person has a different birthday is $365 \times 364 \times \cdots \times (365 - m + 1)$. Denoting by P_m the probability that each person has a different birthday, you have

$$P_m = \frac{365 \times 364 \times \cdots \times (365 - m + 1)}{365^m}.$$

If m is much smaller than 365, the insightful approximation

$$P_m \approx e^{-\frac{1}{2}m(m-1)/365}$$

applies. To see this, write P_m as

$$P_m = 1 \times \left(1 - \frac{1}{365}\right) \times \left(1 - \frac{2}{365}\right) \times \cdots \times \left(1 - \frac{m-1}{365}\right).$$

Next, using the approximation $e^{-x} \approx 1 - x$ for x close to zero and the well-known algebraic formula $1 + 2 + \cdots + n = \frac{1}{2}n(n + 1)$ for $n \geq 1$, you get

$$P_m \approx e^{-1/365} \times e^{-2/365} \times \cdots \times e^{-(m-1)/365} = e^{-(1+2+\cdots+m-1)/365}$$

$$= e^{-\frac{1}{2}m(m-1)/365}.$$

The sought probability that two or more people share a same birthday is one minus the probability that each person has a different birthday. Thus

probability of two or more people sharing a birthday

$$\approx 1 - e^{-\frac{1}{2}m(m-1)/365}.$$

This probability is already more than 50% for $m = 23$ people (the exact value is 0.5073 and the approximate value is 0.5000). The intuitive explanation that the probability of a match is already more than 50% for such a small value as $m = 23$ is that there are $\binom{23}{2} = 253$ combinations of two persons, each combination having a matching probability of $\frac{1}{365}$.

Natural logarithm

The function e^x is strictly increasing on $(-\infty, \infty)$ with $\lim_{x \to -\infty} e^x = 0$ and $\lim_{x \to \infty} e^x = \infty$. As a consequence, the equation $e^y = c$ has a unique solution y for each $c > 0$. This solution as function of c is called the *natural logarithm* and is denoted by $\ln(c)$ for $c > 0$. Thus the natural logarithm is the inverse function of the exponential function. In other words, $\ln(x)$ is the logarithmic function with base e. The natural logarithm can also be defined by the integral

$$\ln(y) = \int_1^y \frac{1}{v} \, dv \quad \text{for } y > 0.$$

This integral representation of $\ln(y)$, which is often used in probability analysis, shows that the derivative of $\ln(y)$ is

$$\frac{d \ln(y)}{dy} = \frac{1}{y} \quad \text{for } y > 0.$$

Geometric and harmonic series

In probability analysis you will often encounter the geometric series. The basic formula for the geometric series is

$$1 + x + x^2 + \cdots = \frac{1}{1-x} \qquad \text{for } |x| < 1,$$

or, shortly, $\sum_{k=0}^{\infty} x^k = \frac{1}{1-x}$ for $|x| < 1$. You can easily verify this result by working out $(1-x)(1 + x + x^2 + \cdots + x^m)$ as $1 - x^{m+1}$. If you take $|x| < 1$ and let m tend to infinity, then x^{m+1} tends to 0. This gives $(1-x)(1 + x + x^2 + \cdots) = 1$ for $|x| < 1$, which verifies the desired result. Differentiating the geometric series term by term and noting that $\frac{1}{1-x}$ has $\frac{1}{(1-x)^2}$ as derivative, you get

$$1 + 2x + 3x^2 + \cdots = \frac{1}{(1-x)^2} \qquad \text{for } |x| < 1$$

or, shortly, $\sum_{k=1}^{\infty} kx^{k-1} = \frac{1}{(1-x)^2}$ for $|x| < 1$. Similarly, you get $\sum_{k=2}^{\infty} k(k-1)x^{k-2} = \frac{2}{(1-x)^3}$ for $|x| < 1$. For the geometric probability model to be met in the next chapters, these formulas lead to

$$\sum_{k=1}^{\infty} k(1-p)^{k-1}p = \frac{1}{p} \quad \text{and} \quad \sum_{k=1}^{\infty} k^2(1-p)^{k-1}p = \frac{2-p}{p^2}.$$

Finally, the partial sum $1 + \frac{1}{2} + \cdots + \frac{1}{n}$ of the harmonic series appears in a variety of probability problems. An insightful approximation is

$$1 + \frac{1}{2} + \cdots + \frac{1}{n} \approx \ln(n) + \gamma + \frac{1}{2n} \qquad \text{for } n \text{ large,}$$

where $\gamma = 0.57721566\ldots$ is the Euler-Mascheroni constant. The approximation is very accurate. The partial sum $\sum_{k=1}^{n} \frac{1}{k}$ increases extremely slowly as n gets larger.[4]

---

[4]The harmonic series $\sum_{k=1}^{\infty} \frac{1}{k}$ has the value ∞. There are many proofs for this celebrated result. The first proof dates back to about 1350 and was given by the philosopher Nicolas Oresme. His argument is ingenious. Oresme simply observed that $\frac{1}{3} + \frac{1}{4} > \frac{2}{4} = \frac{1}{2}$, $\frac{1}{5} + \frac{1}{6} + \frac{1}{7} + \frac{1}{8} > \frac{4}{8} = \frac{1}{2}$, $\frac{1}{9} + \frac{1}{10} + \cdots + \frac{1}{16} > \frac{8}{16} = \frac{1}{2}$, etc. In general, $\frac{1}{r+1} + \frac{1}{r+2} + \cdots + \frac{1}{2r} > \frac{1}{2}$ for any r, showing that $\sum_{k=1}^{n} \frac{1}{k}$ eventually grows beyond any bound as n gets larger. Isn't it a beautiful argument?

Chapter 2
Basics of Probability

Probability is the science of uncertainty and it's everywhere:

- What is the chance of winning the jackpot in the national lottery?

- What is the chance of having some rare disease if tested positive?

- What is the chance that the last person to draw a ticket will be the winner if one prize is raffled among 10 people?

- How many cards would you expect to draw from a standard deck before seeing the first ace?

- What is the expected value of your loss when you are going to bet 50 times on red in roulette?

- What is the expected number of different values that come up when six fair dice are rolled? What is the expected number of rolls of a fair die it takes to see all six sides of the die?

The tools to answer these kinds of questions will be given in this chapter, which aims to familiarize yourself with the most important basic concepts in elementary probability. The standard axioms of probability are introduced, the important properties of probability are derived, the key ideas of conditional probability and Bayesian thinking are covered, and the concepts of random variable, expected value and standard deviation are explained. All this is illustrated with insightful examples and instructive problems.

2.1 Foundation of probability

Approximately four hundred years after the colorful Italian mathematician and physician Gerolamo Cardano (1501–1576) wrote his book *Liber de Ludo Aleae* (*Book on Games of Chance*) and laid a cornerstone for the foundation of the field of probability by introducing the concept of *sample space*, celebrated Russian mathematician Andrey Kolmogorov (1903–1987) cemented that foundation with axioms on which a solid theory can be built.

The sample space of a chance experiment is a set of elements that one-to-one correspond to all of the possible outcomes of the experiment. Here are some examples:

• The experiment is to roll a die once. The sample space can be taken as the set $\{1, 2, \ldots, 6\}$, where the outcome i means that i dots appear on the up face.

• The experiment is to repeatedly roll a die until the first six shows up. The sample space can be taken as the set $\{1, 2, \ldots\}$ of the positive integers. Outcome k indicates that a six appears for the first time on the kth roll.

• The experiment is to measure the time until the first emission of a particle from a radioactive source. The sample space can be taken as the set $(0, \infty)$ of the positive real numbers, where the outcome t indicates that it takes a time t until the first emission of a particle.

In the first example the sample space is a finite set. In the second example the sample space is a so-called countably infinite set, while in the third example the sample space is a so-called uncountable set. In general, a non-finite set is called *countably infinite* if the elements of the set one-to-one correspond to the natural numbers. Not all sets with a non-finite number of elements are countably infinite. The set of all points on a line and the set of all real numbers between 0 and 1 are examples of infinite sets that are not countable. Sets that are neither finite nor countably infinite are called *uncountable*, whereas sets that are either finite or countably infinite are called *countable*.

The idea of Kolmogorov was to consider a sufficiently rich class of subsets of the sample space and to assign a number, $P(A)$, between 0 and 1, to each subset A belonging to this class of subsets. The class of subsets consists of all possible subsets if the sample space is finite or countably infinite, but certain 'weird' subsets must be excluded if the sample space is uncountable. For the probability measure P, three natural postulates are assumed. Denoting by $P(A$ or $B)$ the number assigned to the set of all outcomes belonging to either subset A or subset B or to both, the axioms for a finite sample space are:

Axiom 1. $P(\Omega) = 1$ *for the sample space* Ω.

Axiom 2. $0 \leq P(A) \leq 1$ *for each subset A of* Ω.

Axiom 3. $P(A$ or $B) = P(A) + P(B)$ *if the subsets A and B have no element in common (so-called disjoint subsets).*

Axiom 3 needs a modification if the sample space is non-finite. Then the axiom is $P(A_1$ or A_2 or $\ldots) = \sum_{k=1}^{\infty} P(A_k)$ for pairwise disjoint subsets $A_1, A_2, \ldots$. The sample space endowed with a probability measure P on the class of subsets is called a *probability space*.

If the sample space contains a countable number of elements, it is sufficient to assign a probability $p(\omega)$ to each element ω of the sample space. The probability $P(A)$ that is then assigned to a subset A of the sample space is defined by the sum of the probabilities of the individual elements of set A. That is, in mathematical notation,

$$P(A) = \sum_{\omega \in A} p(\omega).$$

A special case is the case of a finite sample space in which each outcome is equally likely. Then $P(A)$ can be calculated as

$$P(A) = \frac{\text{the number of outcomes belonging to } A}{\text{the total number of outcomes of the sample space}}.$$

This probability model is known as the *Laplace model*, named after the famous French scientist Pierre Simon Laplace (1749–1827), who is sometimes called the 'French Newton'. What is called the Laplace model was first introduced by Gerolamo Cardano in his 16th century book. This probability model was used to solve a main problem in early probability: the probability of not getting a 1 in two rolls of a fair die is $\frac{25}{36}$. Galileo Galilei (1564–1642), one of the greatest scientists of the Renaissance, used the model to explain to the Grand Duke of Tuscany, his benefactor, that it is more likely to get a sum of 10 than a sum of 9 in a single roll of three fair dice (the probabilities are $\frac{27}{216}$ and $\frac{25}{216}$).

In probability language, any subset A of the sample space is called an *event*. It is said that event A occurs if the outcome of the experiment belongs to the set A. The number $P(A)$ is the probability that event A will occur. Any individual outcome is also an event, but events correspond typically to more than one outcome. For example, the sample space of the experiment of a single roll of a die is the set $\{1, 2, 3, 4, 5, 6\}$, where outcome i means that i dots appear on the up face of the die. Then, the subset $A = \{1, 3, 5\}$ represents the event that an odd number shows up. Events A and B are called *mutually exclusive* (or *disjoint*) if they cannot both occur at the same time. In the experiment of tossing a coin four times, the event of three or more heads and the event of two or more tails are disjoint.

The probability measure P does not appear out of thin air, rather you must consciously choose it. Naturally, this must be done in such a way that the axioms are satisfied and the model reflects the reality of the problem at hand in the best possible way. The axioms must hold true not only for the interpretation of probabilities in terms of relative frequencies for a repeatable experiment such as the rolling of a die. They must also remain valid for the Bayesian interpretation of probability as a measure of personal belief in the outcome of a non-repeatable experiment, such as, for example, a horse race. A subjective probability depends on one's knowledge or information about the event in question.

Example 2.1. What is the probability that the largest number obtained in a single roll of two fair dice will be 4?

Solution. The line of thought is made easier if you imagine that one die is blue, and the other is red. The set consisting of the 36 outcomes (i, j) with $i, j = 1, 2, \ldots, 6$ is taken as sample space for the experiment, where i represents the number of points rolled with the blue die, and j represents the number of points rolled with the red die. The dice are fair so that an appropriate probability model is constructed by assigning the same probability $\frac{1}{36}$ to each element of the sample space. This model is an instance of the Laplace model: in order to find the probability of an event, you count the number of favorable outcomes and divide them by the total number of possible outcomes. Let A be the event that the largest number rolled is 4. The event A occurs only if the experiment gives one of the seven outcomes $(1, 4)$, $(2, 4)$, $(3, 4)$, $(4, 4)$, $(4, 3)$, $(4, 2)$ and $(4, 1)$. Hence

$$P(A) = \frac{7}{36}$$

gives the probability of getting 4 as largest number in one roll of two fair dice.

Sample points may be easily incorrectly counted. In his book *Opera Omnia* the German mathematician Gottfried Wilhelm Leibniz (1646–1716) – inventor of differential and integral calculus along with Isaac Newton – made a famous mistake by stating: "with two dice, it is equally likely to roll twelve points than to roll eleven points, because one or the other can be done in only one manner". He argued: two sixes for a sum 12, and a five and a six for a sum 11. However, there are two ways to get a sum 11, as is obvious by imagining that one die is blue and the other is red. Alternatively, you may think of two rolls of a single die rather than a single roll of two dice.

Another psychologically tempting mistake that is sometimes made is to treat sample points as equally likely while this is actually not the case. This mistake can be illustrated with a famous misstep of Jean le Rond d'Alembert (1717–1783), who was one of the foremost intellectuals of his time. D'Alembert made the error to state that the probability of getting heads in no more than two coin tosses is $\frac{2}{3}$ rather than $\frac{3}{4}$. He reasoned as follows: "once heads appears upon the first toss, there is no need for a second toss. The possible outcomes of the game are thus H, TH and TT, and so the required probability is $\frac{2}{3}$". However, these three outcomes are not equally likely, but should be assigned the respective probabilities $\frac{1}{2}$, $\frac{1}{4}$ and $\frac{1}{4}$. The correct answer is $\frac{3}{4}$, as would be immediately clear from the sample space $\{HH, HT, TH, TT\}$ for the experiment of two coin tosses.

Example 2.2. Two desperados, A and B, are playing a game of Russian roulette, using a gun. One of the gun's six cylinders contains a bullet. The desperados take turns pointing the gun at their own heads and pulling the trigger. Desperado A begins. If no fatal shot is fired, they give the cylinder a spin such that it stops at a random chamber, and the game continues with desperado B, and so on. What is the probability that desperado A will fire the fatal shot?

Solution. The set $\{1, 2, \ldots\}$ of the positive integers is taken as sample space for the experiment. Outcome i means that the fatal shot occurs at the ith trial. An appropriate probability model is constructed by assigning the probability $\frac{1}{6}$ to outcome 1, the probability $\frac{5}{6} \times \frac{1}{6}$ to outcome 2, and, the probability p_i to outcome i, where

$$p_i = \frac{5}{6} \times \cdots \times \frac{5}{6} \times \frac{1}{6} = \left(\frac{5}{6}\right)^{i-1} \times \frac{1}{6} \quad \text{for } i = 1, 2, \ldots.$$

Let A be the event that desperado A fires the fatal shot. This event occurs for the outcomes $1, 3, 5, \ldots$. Thus, using the geometric series $\sum_{k=0}^{\infty} x^k = \frac{1}{1-x}$ for $|x| < 1$, you get

$$P(A) = \sum_{k=0}^{\infty} p_{2k+1} = \frac{1}{6} \sum_{k=0}^{\infty} \left(\frac{25}{36}\right)^k = \frac{1}{6} \times \frac{1}{1 - 25/36},$$

and so the probability that desperado A will fire the fatal shot is $\frac{6}{11}$.

Product rule for a compound chance experiment

The chance experiment from Example 2.2 is a so-called *compound chance experiment*. Such an experiment consists of a sequence of elementary subexperiments. In the compound experiment from Example 2.2 the subexperiments are physically independent of each other, that is, the outcome of one subexperiment does not affect the outcome of any other subexperiment. The probabilities to the outcomes of the compound experiment were assigned by taking the product of probabilities of individual outcomes of the subexperiments. This is the only assignment that reflects the physical independence of the subexperiments.

How to assign probabilities to the outcomes of a compound chance experiment when the subexperiments are not physically independent? This is also done by a *product rule*. To explain this rule, consider the experiment of sequentially picking two balls at random from a box containing four red and two blue balls, where the first picked ball is not put back in the box when drawing the second ball. What is the probability of picking at least one blue ball? This experiment is a compound experiment with two physically dependent subexperiments. The sample space of the compound experiment consists of the four ordered pairs (r, r), (r, b), (b, r) and (b, b), where the first component of each pair indicates the color of the first picked ball and the second component the color of the second picked ball. Outcome (r, r) gets assigned the probability $p(r, r) = \frac{4}{6} \times \frac{3}{5} = \frac{2}{5}$. The rationale behind this assignment is that the first ball you pick will be red with probability $\frac{4}{6}$. If the first ball you pick is red, three red and two blue balls remain in the box, in which case the second ball you pick will be red with probability $\frac{3}{5}$. By a same argument, outcome (r, b) gets assigned the probability $p(r, b) = \frac{4}{6} \times \frac{2}{5} = \frac{4}{15}$, $p(b, r) = \frac{2}{6} \times \frac{4}{5} = \frac{4}{15}$ and $p(b, b) = \frac{2}{6} \times \frac{1}{5} = \frac{1}{15}$. This probability model is an adequate representation of the experiment and enables us to answer the question of what the probability of picking at least one blue ball is. This probability is

$$p(r, b) + p(b, r) + p(b, b) = \frac{4}{15} + \frac{4}{15} + \frac{1}{15} = \frac{3}{5}.$$

The basis of the probability model used to obtain this answer was a product rule. This rule will be encountered again in Section 2.3 when discussing conditional probabilities.

It is fun to give also a probability problem with an uncountable sample space.

Example 2.3. The game of franc-carreau was a popular game in eighteenth-century France. In this game, a coin is tossed on a chessboard. The player wins if the coin does not fall on one of the lines of the board. Suppose a coin with a diameter of d is blindly tossed on a large table. The surface of the table is divided into squares whose sides measure a in length, such that $a > d$. What is the probability of the coin falling entirely within the confines of a square?

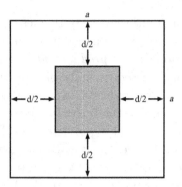

Figure 1: Franc-carreau game

Solution. The trick is to concentrate on the center point of the coin. Take as sample space the square in which this point falls. The meaning of a point in the sample space is that the center of the coin lands on that point. Since the coin lands randomly on the table, the probability that is assigned to each measurable subset A of the sample space is the area of the region A divided by the area of the square. The coin falls entirely within square if and only if the center point of the coin lands on a point in the shaded square in Figure 1. The area of the shaded square is $(a - d)^2$. Therefore

$$P(\text{the coin will fall entirely within a square}) = \frac{(a - d)^2}{a^2}.$$

Complement rule

One of the most useful calculation rules in the field of probability is the complement rule, which states that the probability of a given event occurring can be found by calculating the probability that the event will not occur. These two probabilities sum to 1. The complement rule is often used to find the probability of 'something' occurring at least once. For example, the rule is very helpful to find the probability of at least one six occurring in four rolls of a die and the probability of at least one double six in 24 rolls of two dice, see Problem 2.2 below. This probability problem has an interesting history. The French nobleman Chevalier de Méré was a famous gambler of the 17th century. He frequently offered the bet that he could obtain a six in four rolls or less of a single die and the bet that he could obtain a double six with two dice in 24 rolls or less. The Chevalier believed that the chance of winning the bet was the same in both games (can you explain why the respective chances are not $\frac{4}{6}$ and $\frac{24}{36}$?). In reality, however, he won the first game more often than not. The Chevalier approached the mathematician Blaise Pascal for clarification. This inquiry led to a correspondence between the two famous French mathematicians Blaise Pascal (1623–1662) and Pierre de Fermat (1601–1665).[5] They mathematically clarified the dice problem by simply calculating the chances of *not* rolling a six or double six, see Problem 2.2 below.

The *complement rule* says that, for any event A,

$$\boxed{P(A) = 1 - P(\overline{A}),}$$

where the *complementary event* $\overline{A}$ is defined as the event that A does not occur. A formal proof of this obvious result goes as follows. The events A and $\overline{A}$ are mutually exclusive and together they form the whole sample space. Then, by the Axioms 1 and 3, you have $P(A \text{ or } \overline{A}) = 1$ and $P(A \text{ or } \overline{A}) = P(A) + P(\overline{A})$.

Example 2.4. The game of European roulette is played on a roulette wheel containing 37 slots numbered from $0, 1, \ldots, 36$. In each round of the game, every slot has the same probability of being the one where the roulette ball lands. What is the probability that, during the next 7 rounds, the ball will land in a same slot two or more times?

Solution. Take the set of all possible sequences $(i_1, i_2, \ldots, i_7)$ as sample space, whereby i_k is the number of the slot in which the ball lands in the kth round. The sample space has $37 \times 37 \times \cdots \times 37 = 37^7$ equally likely outcomes. Let A be the event that in some slot the ball lands two or more times. Rather than calculating the probability

---

[5]The 1654 Pascal–Fermat correspondence marks the beginning of modern probability theory. In this correspondence another famous probability problem was solved. Chevalier de Méré had also brought to Pascal's attention the problem of points, in which the question is how the winnings of a game of chance should be divided between two players if the game was ended prematurely. This problem will be discussed in Section 3.1.

$P(A)$, it is simpler to calculate the complementary probability $P(\overline{A})$, where $\overline{A}$ is the event that in none of the slots the ball lands two or more times. The number of outcomes $(i_1, i_2, \ldots, i_7)$ for which the i_k's are different is $37 \times 36 \times \cdots \times 31$. Dividing the number of favorable outcomes by the total number of possible outcomes results in

$$P(\overline{A}) = \frac{37 \times 36 \times \cdots \times 31}{37^7} = 0.5466,$$

or, rather $P(A) = 0.4534$. Would you have imagined that this probability would turn out to be so high?

Problem 2.1. A dog has a litter of four puppies. Use an appropriate sample space to verify that it is more likely that the litter consists of three puppies of the same gender and one of the other than that it consists of two puppies of each gender. (answer: $\frac{8}{16}$ versus $\frac{6}{16}$)

Problem 2.2. Use an appropriate sample space to argue that the probability of getting at least one six in r rolls of a single die is $1 - \frac{5^r}{6^r}$, and the probability of getting at least one double six in r rolls of two dice is $1 - \frac{35^r}{36^r}$. What are the smallest values of r for which the probabilities are more than 0.5? (answer: $r = 4$ and $r = 25$)

Problem 2.3. (a) What is the probability of getting two or more times a same number in a roll of three dice? (answer: $\frac{4}{9}$)
(b) Two dice are rolled. If the biggest number is 1, 2, 3 or 4, player 1 wins; otherwise, player 2. Who has the advantage? (answer: player 2 has a slight advantage)

Problem 2.4. You are randomly dealt four cards from a deck of 52 playing cards with four aces. What is the probability of getting at least one ace? (answer: 0.2813)

Problem 2.5. You have two gift cards, each loaded with 10 free drinks from your favorite coffee shop. Each time you get a drink, you randomly pick one of the cards to pay with. One day it happens for the first time that the waiter can't accept the card because it does not have any drink credits left on it. What is the probability that the other card has also no free drinks on it? (answer: 0.1762)

Problem 2.6. In a game show, a father and his daughter are standing in front of three closed doors, behind which, a car, the key to the car, and a goat are hidden in random order. Each of them can open up to two doors, one at a time, and this must be done out of sight of the other. The daughter is given the task of finding the car, and the father must find the key. Only if both are successful, they get to keep the car. Father and daughter are allowed to discuss a strategy before the game starts. What is an optimal strategy? What is the maximum probability of winning the car? (answer: $\frac{4}{6}$)

Problem 2.7. For non-disjoint sets A and B, the *sum rule* is $P(A \text{ or } B) = P(A) + P(B) - P(A \text{ and } B)$, where '$A$ and B' is the set of outcomes belonging to both A and B. Can you explain this rule? What is the probability of getting an ace or a heart when picking randomly one card from a deck of 52 cards? (answer: $\frac{16}{52}$)

Problem 2.8. An experiment has three possible outcomes O_1, O_2 and O_3 with respective probabilities $p_1 = 0.10$, $p_2 = 0.15$, and $p_3 = 0.75$. What is the probability that outcome O_1 will appear before O_2 if the experiment is done repeatedly? (answer: $\frac{p_1}{p_1 + p_2} = 0.40$)

Problem 2.9. Two people have agreed to meet at the train station. Independently of one other, each person is to appear at a random moment between 12 p.m and 1 p.m. What is the probability that they will meet within 10 minutes of each other? (answer: $\frac{11}{36}$)

2.2 The concept of conditional probability

The concept of conditional probability lies at the heart of probability theory. It is an intuitive concept. To illustrate this, most people reason as follows to find the probability of getting two aces when two cards are selected at random in succession from an ordinary deck of 52 cards. The probability of getting an ace on the first card is $\frac{4}{52}$. Given that one ace is gone from the deck, the probability of getting an ace on the second card is $\frac{3}{51}$. Therefore

$$P(\text{the first two cards are aces}) = \frac{4}{52} \times \frac{3}{51}.$$

What is applied here is the *product rule* for probabilities:

$$P(A \text{ and } B) = P(A)P(B \mid A),$$

where $P(A$ and $B)$ stands for the probability that both event A ('the first card is an ace') and event B ('the second card is an ace') will occur, $P(B \mid A)$ is the notation for the conditional probability that event B will occur given that event A has occurred.[6] In words, the unconditional probability that both event A and event B will occur is equal to the unconditional probability that event A will occur times the conditional probability that event B will occur given that event A has occurred. This is one of the most useful rules in probability.

Example 2.5. Someone is looking to rent an apartment on the top floor of a certain building. The person gets wind of the fact that two apartments in the building have been vacated, and are up for rent. The building has seven floors, with eight apartments per floor. What is the probability of having a vacant apartment on the top floor?

Solution. There are two possible approaches to solving this problem. In both, the complement rule is applied. This means that, instead of calculating the probability in question, you calculate the complementary probability of no top floor apartment being available. Subtracting this probability from 1 gives the probability of having a vacant apartment on the top floor.

---

[6]In fact, the other way around, $P(B \mid A)$ is defined as the ratio of $P(A$ and $B)$ and $P(A)$ if $P(A) > 0$. This definition can be motivated as follows. Suppose that n physically independent repetitions of a chance experiment are done under the same conditions. Let r be the number of times that event A occurs simultaneously with event B and s be the number of times that event A occurs but not event B. The frequency at which event B occurs in the cases that event A has occurred is equal to $\frac{r}{r+s}$. The frequency at which both event A and event B occur is $\frac{r}{n}$, and the frequency at which event A occurs is $\frac{r+s}{n}$. The ratio of these frequencies is $\frac{r}{r+s}$. This ratio is exactly the frequency at which event B occurs in the cases that event A has occurred. This explains the definition of $P(B \mid A)$.

The conditional probability $P(B \mid A)$ is in fact a probability measure on a reduced sample space. For example, suppose a blue and red die are rolled and you get the information that there is a six among the two outcomes. Then the reduced sample space of the experiment is $\{(1,6), \ldots, (5,6), (6,1), \ldots, (6,6)\}$, where outcome (i,j) means that the blue and red die show up i and j points.

Approach 1: This approach is based on counting and requires the specification of a sample space. The elements of the sample space are all possible combinations of two of the 56 apartments. The total number of possible combinations is $\binom{56}{2} = 1\,540$, whereas the number of possible combinations without a vacant apartment on the top floor is $\binom{48}{2} = 1\,128$. Then, taking the ratio of all favorable combinations and the total number of combinations,

$$P(\text{no apartment is vacant on the top floor}) = \frac{1\,128}{1\,540} = 0.7325.$$

Approach 2: The second approach is based on conditional probabilities. Imagine that the two available apartments were vacated one after the other. Then, let A be the event that the first vacant apartment is not located on the top floor and B be the event that the second vacant apartment is not located on the top floor. Then $P(A) = \frac{48}{56}$ and $P(B \mid A) = \frac{47}{55}$. Next, by the product rule, you find again the value 0.7325 for the probability that no top floor apartment is available:

$$P(A \text{ and } B) = P(A)P(B \mid A) = \frac{48}{56} \times \frac{47}{55} = 0.7325.$$

Example 2.6. Three boys and three girls are planning a dinner party. They agree that two of them will do the washing up, and they draw lots to determine which two it will be. What is the probability that two boys will wind up doing the washing up?

Solution. A useful solution strategy in probability is to see whether your problem is not the same as another problem, for which the solution is more obvious. This is the situation, here. The sought-after probability is the same as the probability of getting two red balls, when blindly choosing two balls from a bowl containing three red and three blue balls. If A represents the event that the first ball chosen is red, and B represents the event that the second ball chosen is red, then the sought-after probability is equal to $P(A \text{ and } B)$. Thus, using the basic formula $P(A \text{ and } B) = P(A)P(B \mid A)$, you find that $P(\text{two boys will do the washing up}) = \frac{3}{6} \times \frac{2}{5} = \frac{1}{5}$.

An obvious extension of the product formula is

$$P(A_1 \text{ and } A_2 \text{ and } \ldots \text{ and } A_n)$$
$$= P(A_1) \times P(A_2 \mid A_1) \times \cdots \times P(A_n \mid A_1 \text{ and } A_2 \text{ and} \ldots \text{ and } A_{n-1}).$$

This useful extension is illustrated with the following example.

Example 2.7. What is the probability that you must pick five or more cards from a shuffled deck of 52 cards before getting an ace?

Solution. Noting that the sought-after probability is nothing else than the probability of getting no ace among the first four picked cards, let A_i be the event that the ith picked card is not an ace for $i = 1, \ldots, 4$. The probability $P(A_1 \text{ and } A_2 \text{ and } A_3 \text{ and } A_4)$ is the probability that five or more cards are needed to get an ace. This probability is calculated from the extended product formula with $n = 4$ and has the value $\frac{48}{52} \times \frac{47}{51} \times \frac{46}{50} \times \frac{45}{49} = 0.7187$.

An alternative calculation is as follows: let E_k be the event that the first $k - 1$ cards are non-aces and F_k be the event that the kth card is an ace. Then, the probability p_k of getting the first ace at the kth pick is $P(E_k \text{ and } F_k) = P(E_k)P(F_k \mid E_k)$. Verify yourselves

$$p_k = \frac{\binom{48}{k-1}}{\binom{52}{k-1}} \times \frac{4}{52 - (k-1)} \qquad \text{for } k = 1, 2, \ldots, 49,$$

which gives $\sum_{k=5}^{49} p_k = 0.7187$. It never hurts to solve a problem in different ways. It allows you to double check your answer.

The foregoing examples show that when you use an approach based on conditional probabilities to solve the problem, you usually go straight to work without first defining a sample space. The counting approach, however, does require the specification of a sample space. If both approaches are possible for a given problem, then the approach based on conditional probabilities will, in general, be simpler than the counting approach.

For events A and B with nonzero probabilities, the formula

$$P(B \mid A) = \frac{P(A \text{ and } B)}{P(A)}$$

quantifies how the original probability $P(B)$ changes when new information becomes available. If $P(B \mid A) = P(B)$, then the events A and B are said to be *independent*. An equivalent definition of independence is $P(A \text{ and } B) = P(A)P(B)$. The concept of independence will be further explored in Section 2.7.

Beginning students sometimes think that independent events A and B with nonzero probabilities are disjoint. This is not true. The explanation is that $P(A \text{ and } B) = 0$ if A and B are disjoint, whereas $P(A \text{ and } B) = P(A)P(B) > 0$ if A and B are independent.

Problem 2.10. Five friends are sitting at a table in a restaurant. Two of them order white wine and the other three order red wine. The waiter has forgotten who ordered what and puts the drinks in random order before the five persons. What is the probability that each person gets the correct drink? (answer: $\frac{1}{10}$)

Problem 2.11. A bag contains 14 red cards and 7 black cards. You pick two cards at random from the bag. Verify that it is more likely to pick one red and one black card rather than two red cards. (answer: the probabilities are $\frac{14}{30}$ and $\frac{13}{30}$)

Problem 2.12. Someone has rolled two dice out of your sight. You ask this person to answer "yes or no" on the question whether there is a six among the two rolls. He truthfully answers "yes." What is the probability that two sixes have been rolled? (answer: $\frac{1}{11}$)

Problem 2.13. A prize is raffled among 10 people. In a previously agreed order, each of them draws a lottery ticket from a bowl with 10 tickets, including one winning ticket. What is the probability that the kth person in the row will win the prize? (answer: $\frac{1}{10}$ for all k)

Problem 2.14. In a variation of the hilarious TV-show game of Egg Russian roulette, two participants are shown an egg box with four boiled eggs and two raw eggs in random order. They take turns taking an egg and smashing it upon their heads. What is the probability that the person who begins smashes the first raw egg? (answer: 0.6) How does this probability change when there are five boiled eggs and one raw egg? (answer: 0.5)

Problem 2.15. Four British teams are among the eight teams that have reached the quarter-finals of the Champions League soccer. What is the probability that the four British teams will avoid each other in the quarter-finals draw if the eight teams are paired randomly? (answer: $\frac{8}{35}$) *Hint*: think of a bowl containing four red and four blue balls, where you remove each time two randomly chosen balls from the bowl. What is the probability that you remove each time a red and a blue ball? Solving a probability problem becomes often simpler by casting the problem into an equivalent form.

Problem 2.16. If you pick at random two children from the Johnson family, the chances are 50% that both children have blue eyes. How many children does the Johnson family have and how many of them have blue eyes? (answer: 4 and 3)

Problem 2.17. Your friend shakes thoroughly two dice in a dice-box. He then looks into the dice-box. Your friend is honest and always tells you if he sees a six in which case he bets with even odds that both dice show an even number. Is the game favorable to you? (answer: yes, your probability of winning is $\frac{6}{11}$)

2.3 The law of conditional probability

Suppose that a closed box contains one ball. This ball is white. An extra ball is added to the box and the added ball is white or red with equal chances. Next one ball is blindly removed from the box. What is the probability that the removed ball is white? A natural reasoning is as follows. The probability of removing a white ball is 1 if a white ball has been added to the box and is $\frac{1}{2}$ if a red ball has been added to the box. It is intuitively reasonable to average these conditional probabilities over the probability that a white ball has been added and the probability that a red ball has been added. The latter two probabilities are both equal to $\frac{1}{2}$. Therefore

$$P(\text{the removed ball is white}) = 1 \times \frac{1}{2} + \frac{1}{2} \times \frac{1}{2} = \frac{3}{4}.$$

This is an application of the *law of conditional probability*. This law calculates a probability $P(A)$ with the help of appropriately chosen conditioning events B_1 and B_2. These events should be such that event A can occur only if one of the events B_1 and B_2 have occurred, and the events B_1 and B_2 must be disjoint. Then, $P(A)$ can be calculated as

$$P(A) = P(A \mid B_1)P(B_1) + P(A \mid B_2)P(B_2).$$

This is a very useful rule to calculate probabilities.[7] The extension of the rule to more than two conditioning events B_i is obvious. In general, the choice of the conditioning events is self-evident. In the above example of the two balls, the conditioning event B_1 is the event that a white ball has been added to the box and B_2 is the event that a red ball has been added.

Example 2.8. A drunkard removes two randomly chosen letters of the message HAPPY HOUR that is attached on a billboard outside a pub. His drunk friend puts the two letters back in a random order. What is the probability that HAPPY HOUR appears again?

Solution. Let A be the event that the message HAPPY HOUR appears again. In order to calculate $P(A)$, it is obvious to condition on the two events B_1 and B_2, where B_1 is the event that two identical letters have been removed and B_2 is the event that two different letters have been removed. In order to apply the law of conditional probability, you need to know the probabilities $P(B_1)$, $P(B_2)$, $P(A \mid B_1)$ and $P(A \mid B_2)$. The latter two probabilities are easy: $P(A \mid B_1) = 1$ and $P(A \mid B_2) = \frac{1}{2}$. The probabilities $P(B_1)$ and $P(B_2)$ require some more thought. It suffices to determine $P(B_1)$, because $P(B_2) = 1 - P(B_1)$. The probability $P(B_1)$ is the sum of the probability that the drunkard has removed the two H's and the probability that the drunkard has removed the P's. Each of the latter two probabilities is equal to $\frac{2}{9} \times \frac{1}{8} = \frac{1}{36}$, by the product rule. Thus $P(B_1) = \frac{1}{18}$ and $P(B_2) = \frac{17}{18}$. By the law of conditional probability,

$$P(A) = 1 \times \frac{1}{18} + \frac{1}{2} \times \frac{17}{18} = \frac{19}{36},$$

---

[7]The proof is simple. Since A can only occur if one the events B_1 or B_2 has occurred and the events B_1 and B_2 are disjoint, $P(A) = P(A \text{ and } B_1) + P(A \text{ and } B_2)$ (by Axiom 3 in Section 2.1). The product rule next leads to $P(A) = P(A \mid B_1)P(B_1) + P(A \mid B_2)P(B_2)$.

and so the probability that the message appears again is $\frac{19}{36}$.

Problem 2.18. Michael arrives home on time with probability 0.8. If Michael does not arrive home on time, the probability that his dinner is burnt is 0.5; otherwise, this probability is 0.15. What is the probability that Michael's dinner will be burnt? (answer: 0.22)

Problem 2.19. Your friend has chosen at random a card from a standard deck of 52 cards, but keeps this card concealed. You have to guess which of the 52 cards it is. Before doing so, you can ask your friend either the question whether the chosen card is red or the question whether the card is the ace of spades. Your friend will answer truthfully. What question would you ask? (answer: the probability of a correct guess is $\frac{1}{26}$ in both cases)

Problem 2.20. One fish is contained in an opaque fishbowl. The fish is equally likely to be a piranha or a goldfish. A sushi lover throws a piranha into the fishbowl alongside the other fish. Then, immediately, before either fish can devour the other, one of the fish is blindly removed from the fishbowl. The removed fish appears to be a piranha. What is the probability that the fish that was originally in the bowl by itself was a piranha? (answer: $\frac{2}{3}$)

Problem 2.21. In a lotto 6/45 draw, six different numbers are randomly drawn from the numbers 1 to 45. You win the jackpot if you have predicted correctly all six numbers drawn. If you have exactly two numbers correctly predicted on a ticket, you get a free ticket for the next lotto draw. What is the probability that you will ever win the jackpot when you buy for once and for all a single ticket for playing in the lotto? (answer: 1.447×10^{-7})

Problem 2.22. On the TV show 'Deal or No Deal', you are faced with 26 briefcases in which various amounts of money have been placed including the amounts \$2 500 000 and \$1 000 000. You first choose one case. This case is 'yours' and is kept out of play until the very end of the game. Then you play the game and in each round you open several cases. What is the probability that the cases with \$2 500 000 and \$1 000 000 will be still unopened at the end of the game when you are going to open 20 cases? (answer: $\frac{3}{65}$)

2.4 Bayesian probability

The Bayesian view of probability is interwoven with conditional probability. Bayes' formula, which is nothing else than logical thinking, is the most important rule in Bayesian probability.[8] To introduce this rule, consider again Problem 2.18. In this problem, Michael finds his dinner burnt (event A) with probability 0.5 if he does not arrive home on time (event B). That is, $P(A \mid B) = 0.5$. Suppose you are asked to give the probability $P(B \mid A)$, being the conditional probability that Michael did not arrive home on time given that his dinner is burnt. In fact you are asked to reason back from effect to cause. Then, you are in the area of Bayesian probability. The *basic form of Bayes' rule* is

$$P(B \mid A) = \frac{P(B)P(A \mid B)}{P(A)}$$

for any two events A and B with $P(A) > 0$. The derivation of this formula is strikingly simple. The basic form of Bayes' rule follows directly from the definition of conditional probability:

$$P(B \mid A) = \frac{P(B \text{ and } A)}{P(A)} = \frac{P(B)P(A \mid B)}{P(A)}.$$

In Problem 2.18, the values of $P(B)$ and $P(A \mid B)$ were given as 0.2 and 0.5, and the value of $P(A)$ was calculated as 0.22. Thus the probability that Michael did not arrive home on time given that his dinner is burnt is equal to

$$P(B \mid A) = \frac{0.2 \times 0.5}{0.22} = \frac{5}{11}.$$

You see that Bayes' rule enables you to reason back from effect to cause in terms of probabilities. Many interesting queries are matters

---

[8]This formula is named after the English clergyman Thomas Bayes (1702–1762) who derived a special case of the formula. The formula in its general form was first written down by Pierre Simon Laplace (1749–1827). The famous British scientist Sir Harold Jeffreys (1891–1989) once stated that Bayes' formula is to the theory of probability what the Pythagorean theorem is to geometry.

of statistical inference, where the aim is to reason "backwards" from observed effects to unknown causes. In medical diagnosis, for example, the physician records a set of symptoms and must identify the underlying disease. Bayes' rule is the answer to such questions.

There are various versions for Bayes' rule. The most insightful version is the Bayes' rule in odds form. This version is mostly used in practice. Before stating Bayes' rule in odds form, the concept of odds will be discussed. Let G be any event that will occur with probability p, and so event G will not occur with probability $1 - p$. Then the *odds* of event G are defined by:

$$o(G) = \frac{p}{1 - p}.$$

Conversely, the odds $o(G)$ of an event G determines $p = P(G)$ as

$$p = \frac{o(G)}{1 + o(G)}.$$

For example, an event G with probability $\frac{2}{3}$ has odds 2 (it is often said the odds are 2:1 in favor of event G), while an event with odds $\frac{2}{9}$ (odds are 2:9) has a probability $\frac{2}{11}$ of occurring.

Bayes' rule in odds form will be formulated in terms of events H (hypothesis) and E (evidence) rather than events A and B. Also, the standard notation $\overline{H}$ is used for the event that event H does not occur. Then, *Bayes' rule in odds form* reads as[9]

$$\frac{P(H \mid E)}{P(\overline{H} \mid E)} = \frac{P(H)}{P(\overline{H})} \times \frac{P(E \mid H)}{P(E \mid \overline{H})}.$$

What does this formula say and how to use it? This is easiest explained with the help of an example.

---

[9]This rule is obtained as follows. The basic form of the formula of Bayes gives that $P(H \mid E) = P(H)P(E \mid H)/P(E)$ and $P(\overline{H} \mid E) = P(\overline{H})P(E \mid \overline{H})/P(E)$. Taking the ratio of these two expressions, $P(E)$ cancels out and you get Bayes' rule in odds form. The derivation shows that the formula is also true when $\overline{H}$ would not be the complement of H. That is, for any two hypotheses H_1 and H_2, the general Bayes formula $\frac{P(H_1|E)}{P(H_2|E)} = \frac{P(H_1)}{P(H_2)} \times \frac{P(E|H_1)}{P(E|H_2)}$ applies.

Suppose that a team of divers believes that a sought-after wreck will be in a certain sea area with a probability of $p = 0.4$. A search in that area will detect the wreck with a probability of $d = 0.9$ if it is there. What is the revised probability of the wreck being in the area when the area is searched and no wreck is found? To answer this question, let hypothesis H be the event that the wreck is in the area in question and thus $\overline{H}$ is the event that the wreck is not in that area. Before the search takes place, the events H and $\overline{H}$ have probabilities $P(H) = 0.4$ and $P(\overline{H}) = 0.6$. These probabilities are called *prior probabilities*. The ratio of $P(H)$ and $P(\overline{H})$ is the *prior odds* of hypothesis H. These odds will change if additional information becomes available. Denote by evidence E the event that the search is not successful. The probability $P(E \mid H) = 1 - 0.9 = 0.1$. Obviously, the probability $P(E \mid \overline{H}) = 1$. The ratio of $P(E \mid H)$ and $P(E \mid \overline{H})$ is called the *likelihood ratio* or *Bayes factor*. It will be clear that the evidence supports the hypothesis if the likelihood ratio is greater than 1 and supports the negation of the hypothesis if the likelihood ratio is less than 1. Once the prior odds and the likelihood factor have been determined, Bayes' rule in odds form can be applied to calculate the *posterior odds* of hypothesis H. These posterior odds are $o(H \mid E) = P(H \mid E)/P(\overline{H} \mid E)$ and so the *posterior probability* of hypothesis H is given by:

$$\boxed{P(H \mid E) = \frac{o(H \mid E)}{1 + o(H \mid E)}.}$$

The posterior probability $P(H \mid E)$ gives the updated value of the probability that hypothesis H is true after that additional information has become available through the evidence event E. Bayesian updating – revising an estimate when new information is available – is a key concept in statistics and data science.

For the search of the wreck, Bayes' rule in odds form gives

$$\frac{P(H \mid E)}{P(\overline{H} \mid E)} = \frac{0.4}{0.6} \times \frac{0.1}{1} = \frac{1}{15}.$$

Thus the posterior probability of hypothesis H is

$$P(H \mid E) = \frac{1/15}{1 + 1/15} = \frac{1}{16}.$$

This is the revised value of the probability that the wreck is in the area in question after the futile search.

In words, Bayes' rule in odds form reads as:

$$\boxed{posterior\ odds = prior\ odds \times likelihood\ ratio.}$$

It is emphasized that the prior odds of hypothesis H refer to the situation *before* the occurrence of evidence event E, whereas the posterior odds refer to the situation *after* the occurrence of event E.

Example 2.9. An athlete selected by lot has to go to the doping control. On average, 7 out of 100 athletes use doping. The doping test gives a positive result with a probability of 96% if the athlete has used doping and with a probability of 5% if the athlete has not used doping. Suppose that the athlete gets a negative test result. What is the probability that the athlete has nevertheless used doping?

Solution. Let the hypothesis H be the event that the athlete has used doping. The prior probabilities are $P(H) = 0.07$ and $P(\overline{H}) = 0.93$. Let E be the event that the athlete has a negative test result. Then, $P(E \mid H) = 0.04$ and $P(E \mid \overline{H}) = 0.95$. The posterior odds of the hypothesis H are

$$\frac{P(H \mid E)}{P(\overline{H} \mid E)} = \frac{0.07}{0.93} \times \frac{0.04}{0.95} = 0.003169.$$

Thus the revised value of the probability of doping use notwithstanding a negative test result is

$$P(H \mid E) = \frac{0.003169}{1 + 0.003169} = 0.003159,$$

or, rather about 0.32%. A very small probability indeed.

The posterior probability of 0.32% can also be calculated without using conditional probabilities and Bayes' rule. The alternative calculation is based on the method of *expected frequencies*. This method is also easy to understand by the layman. Imagine a very large number of athletes that are selected by lot for the doping control, say 10 000 athletes. On average 700 of these athletes have used doping and on average 9 300 athletes have not used doping. Of these 700 athletes, $700 \times 0.04 = 28$ athletes test negative on average, whereas $9\,300 \times 0.95 = 8\,835$ athletes of the other 9 300 athletes test negative on average. Thus a total of $28 + 8\,835 = 8\,863$ athletes test negative and among those 8 863 athletes there are 28 doping users. Therefore the probability that an athlete has used doping notwithstanding a negative test result is $\frac{28}{8\,863} = 0.003159$. The same probability as found with Bayes' rule. A similar reasoning shows that the probability that an athlete with a positive test result has not used doping is $\frac{465}{465+672} = 0.40987$ (verify!).

The following problems ask you to apply Bayes' rule in odds form. You should first identify the hypothesis H and the evidence E.

Problem 2.23. An oil explorer performs a seismic test to determine whether oil is likely to be found in a certain area. The probability that the test indicates the presence of oil is 90% if oil is indeed present in the test area, while the probability of a false positive is 15% if no oil is present in the test area. Before the test is done, the explorer believes that the probability of presence of oil in the test area is 40%. What is the revised probability of oil being present in the test area given that the test is positive? (answer: 0.8).

Problem 2.24. Consider Problem 2.18 again. What is the probability that Michael arrived at home on time given that he did not find his dinner burnt? (answer: $\frac{34}{39}$)

Problem 2.25. On the island of liars each inhabitant lies with probability $\frac{2}{3}$. You overhear an inhabitant making a statement. Next you ask another inhabitant whether the inhabitant you overheard spoke truthfully. What is the probability that the inhabitant you overheard indeed spoke truthfully given that the other inhabitant says so? (answer: $\frac{1}{5}$)

Problem 2.26. You have two symmetric dice in your pocket. One die is a standard die and the other die has each of the three numbers 2, 4, and 6 twice on its faces. You random pick one die from your pocket without looking. Someone else rolls this die and informs you that a 6 has shown up. What is the revised value of the probability that you have picked the standard die? How does this probability change if the die is rolled a second time and a 6 appears again? (answer: $\frac{1}{3}$ and $\frac{1}{5}$)

Problem 2.27. Someone visits the doctor because he fears having a very rare disease. This disease occurs in only 0.1% of the population. The doctor proposes a test that correctly identifies 99% of the people who have the disease and only incorrectly identifies 1% of the people who don't have the disease. Suppose the person in question has tested positive. What is the probability that he actually has the disease? (answer: 9.02%) How does the answer change if a second independent test is also positive? (answer: 90.75%)

The case of Sally Clark: a miscarriage of justice

Sally Clark was arrested in 1999 after her second child, who was a few months old, died, ostensibly by cot death, just as her first child had died a year earlier. She was accused of suffocating both children. During the trial the prosecutor called a famous pediatrician as an expert. He stated that the chance of cot death of a child was about 1 in 8 543 and stated that the chance of two cot deaths in the same family was $\left(\frac{1}{8513}\right)^2$, or, about 1 in 73 million. The prosecutor argued that, beyond any reasonable doubt, Sally Clark was guilty of murdering her two children, and the jury sentenced her to life imprisonment, though there was no other evidence that Sally Clark had killed her two children. This is a classic example of the 'prosecutor's fallacy'. The probability of innocence given the death of the two children – the probability that matters – is confused with the tiny probability that in the same family two infant children will die of sudden infant death syndrome.

The conviction of Sally Clark led to great controversy and several leading British statisticians threw themselves into the case. The statisticians came up with various estimates for Sally Clark's chance of innocence and all these estimates showed that the condemnation of her was not beyond reasonable doubt. The formula of Bayes was the basis of the calculations of the statisticians. How did this work? Let H be the event that Sally Clark is guilty and the evidence E be the event that both of her children died in the first few months of their lives. The probability that matters is the conditional probability $P(H \mid E)$. To get this probability, you need prior probabilities $P(H)$ and $P(\overline{H})$ together with likelihood ratio $P(E \mid H)/P(E \mid \overline{H})$. The assumption is made that murder by the mother (hypothesis H) and cot death (hypothesis $\overline{H}$) are the only two possibilities for the death of the two children. Of course, $P(E \mid H) = 1$. The pediatrician called as expert gave the estimate $\frac{1}{8\,543} \times \frac{1}{8\,543}$ for $P(E \mid \overline{H})$, but this estimate assumes independence between both deaths. However, a cot death in a family increases the likelihood that a subsequent birth in the family will also die of cot death. In an article in the British Medical Journal it was made plausible that a factor of 5 applies to the increased chance. Thus the probability $P(E \mid \overline{H})$ is estimated by

$$P(E \mid \overline{H}) = \frac{1}{8\,543} \times \frac{5}{8\,543} \approx 6.85 \times 10^{-8},$$

or, about 1 in 14.8 million, which is still a very small probability. However, this probability should be weighed with the very small prior probability that a mother will kill both of her children at the beginning of their first year of life by suffocation. How do you get a good estimate for the prior probability $P(H)$? This is not simple. However, on the basis of statistical data, an upper bound for the prior probability $P(H)$ can be estimated. Instead of asking how often mothers in a family like the Clarks kill their first two children in their first year of life, the question can be answered on how often mothers kill one or more of their children of any age. Data are available in the U.S. Statistics give about 100 cases per year in the U.S. In the U.S. there are about 120 million adult women and about half of them have children, so about 1 in 0.6 million American women murder one or more of their children. The frequency of murders in America is about 4 times as large as in England. This leads to the estimate that about 1 in 2.4 million women in England kill one or more of their children. This is, of course, an overestimate of the prior probability $P(H)$. If you nevertheless take

$$P(H) = \frac{1}{2.4 \times 10^6},$$

then you find with Bayes' rule in odds form that

$$\frac{P(H \mid E)}{1 - P(H \mid E)} \approx \frac{1/(2.4 \times 10^{-6})}{1 - 1/(2.4 \times 10^{-6})} \times \frac{1}{6.85 \times 10^{-8}} \approx 6.08.$$

Thus the posterior probability of Sally Clark's guilt is given by

$$P(H \mid E) \approx \frac{6.08}{1 + 6.08} = 0.859.$$

Therefore the probability of Sally Clark's innocence is estimated by 0.141. In fact, this is an underestimate, since 0.859 is an overestimate of the probability of Sally Clark's guilt.[10] So a probability of 14.1% or more is a reasonable estimate for the probability of Sally Clark's innocence. This is of course no base for a conviction when there is no other evidence. Despite the arguments that statisticians presented, Sally Clark lost the appeal against her conviction. But in 2003 she was acquitted after it came out that her second child had a bacterial infection in the brain at the time of his death, a fact that was withheld from the defense in the earlier trial. The tragic event surrounding Sally Clark is similar with the miscarriage of justice that took place in the Netherlands around the nurse Lucia de Berk who was wrongly accused of murdering a number of her patients who died during her night shifts.

People v. Collins

An older famous example of the prosecutor's fallacy is People v. Collins, an American robbery trial. On June 18, 1964, Juanita Brooks was attacked in an alley near to her home in Los Angeles and her purse stolen. A witness reported that a white woman running from the scene was blond, had a pony tail, and fled from the scene in a yellow car driven by a black man with a beard and a mustache. Police arrested a couple, Janet and Mark Collins, which fit the description. Unfortunately for the prosecutor neither Juanita Brooks nor the witness could make a positive identification of either of the defendants. At the trial, following testimony by a college mathematics instructor, the prosecutor provided the following probabilities of occurrence of the reported characteristics: girl with blond hair $\frac{1}{3}$, girl with ponytail $\frac{1}{10}$, yellow car $\frac{1}{10}$, man with mustache $\frac{1}{4}$, black man with beard $\frac{1}{10}$, and interracial couple in car $\frac{1}{1\,000}$. Although the instructor had told the prosecutor that these individual

---

[10] An overestimate of the prior $P(H)$ always results in an overestimate of the posterior $P(H \mid E)$. This intuitive result can be deduced from Bayes' formula, using the fact that $a/(1 - a) < b/(1 - b)$ if and only if $a < b$ when $0 < a, b < 1$.

probabilities could not simply be multiplied by each other, the prosecutor multiplied them to arrive at a joint probability of

$$\frac{1}{3} \times \frac{1}{10} \times \frac{1}{10} \times \frac{1}{4} \times \frac{1}{10} \times \frac{1}{1\,000} = \frac{1}{12\,000\,000}.$$

In his summation, the prosecutor emphasized the extreme unlikelihood that a couple other than the defendants had all these characteristics. Impressed by the long odds, the jury convicted the couple for second-degree robbery. But did they make the right decision? The answer is no! The fundamental flaw in the prosecutor's reasoning was to equate the probability of innocence of the Collins couple with the probability that a randomly chosen couple would have the six characteristics in question. Moreover, the individual probabilities of the characteristics were unfounded and, by dependence between the characteristics (beards and mustaches are not independent events), it was wrong to multiply them to arrive at a probability of 1 in 12 million. Defense appealed and the California Supreme Court reversed the conviction, very critical of the statistical reasoning used and the way it was put to the jury.

O.J. Simpson trial

A remarkable example of confusing conditional probabilities happened in the O.J. Simpson trial. This trial, regarded by many as 'the trial of the century', dominated the news for more than a year and was broadcast on television. In 1994, O.J. Simpson, an actor and former American football star, was accused of murdering his ex-wife, Nicole Brown. The trial started with prosecution proving that O.J. Simpson has a history of physical violence against his ex-wife Nicole. The famous lawyer Alan Dershowitz countered for the defense. Dershowitz argued: "only about one in 2 500 men who batter their domestic partners go on to murder them, so the fact that O.J. Simpson battered his wife is irrelevant to the case". This was a clever trick to fool the non-mathematical jury. The defense lawyer confused the jury with asking the question: what's the probability of a man murdering his partner given that he previously battered her? This conditional probability is indeed about 0.04% but it is not the right statistic. It ignores the crucial fact that Nicole Brown was actually murdered. The real question is: what's the probability that a man murdered his wife given that he previously battered her and she was murdered? That conditional probability turns out to be very far from 0.04%. It can be approximated by Bayes' rule. Instead, let's use the expected frequency approach. Imagine a sample of 100 000 battered women. According to Dershowitz's number of 1 in 2 500, you can expect about 40 of these women to be murdered by their violent partners in a given year. Let's take as rough estimate that an additional 5 of the 100 000 battered women will be killed by someone else, on the basis of the fact that the murder rate for *all* women in the United States at the time of the trial was about 1 in 20 000 per year. Then the conditional probability that a man murdered his wife given that he previously battered her and she was murdered can be approximated by

$$\frac{40}{40+5} = \frac{8}{9},$$

or, about 89%. This probability is much and much larger than the probability of 0.04% in the defense lawyer's argument. The estimated probability of 89% shows that the fact that O.J. Simpson had physically abused his wife in the past was certainly very relevant to the case. This probability should not be confused with the probability that O.J. Simpson did it. Many other factors also determine this probability.

Bayesian vs. classical statistics

Bayesian probability is the basis of Bayesian statistics. This field of statistics is very different from the field of classical statistics dealing with hypothesis-testing. Imagine that a new medication is being tested on a given number of patients, and that it appears to be effective for a number of them. You want to know if this means that the medication works. In classical statistics, you would start with the assumption that mere fluke is the cause of the test results (this is called the null hypothesis). The null hypothesis is then tested using the so-called p-value, being the probability of getting data that are *at least* as good as the observed data if the null hypothesis would be true. If the p-value is below some threshold value – the value 0.05 is often used as a cut-off value – the null hypothesis is rejected and it is assumed that the medication is effective (it is said that the findings are 'statistically significant'). The p-value, however, does not tell you what the probability is that the new medication is not effective. And this is, in fact, the probability you really want to know. Scientific studies have shown that the p-value can give a highly distorted picture of this probability. The probability that the medication is not effective can be considerably larger than the p-value. As a consequence, you must be careful with drawing a conclusion when the p-value is just below 0.05; the test $p < 0.05$ is not a litmus test. As such the test was never intended, but it was meant as a signal to investigate matters further. In the case that the p-value is extremely small and the study is carefully designed, then further investigations are not necessary. Physicists reported the 'discovery' of the Higgs boson in 2012 after a statistical analysis of the data showed that they had attained a confidence level of about a one-in-3.5 million probability that the experimental results would have been obtained if there were no Higgs particle. Again, in all clarity, this p-value is not the probability that the Higgs boson doesn't exist. Another extremely small p-value occurred in the famous 1954 Salk vaccine polio study. In this statistical study, a double-blind study was conducted with two groups of 200 000 children each. A total of 142 children in the placebo group developed polio and 42 children in the vaccine group. The p-value was on the order of 10^{-9} and removed any doubts about the vaccine's efficacy.

Bayesian statistics enables you to give a judgment about the probability that the medication works. The judgment uses the generic formula:

$$p(\theta \mid data) = \frac{p(data \mid \theta)p(\theta)}{p(data)}.$$

What is the meaning of the elements of this formula? This formula determines the posterior probability distribution $p(\theta \mid data)$ of an unknown parameter θ. For example, you can think of θ as the percentage of people for whom a new medication is working. To estimate the posterior distribution, you need data from test experiments. Before the tests are done, you should specify a prior probability distribution $p(\theta)$ on the parameter θ. The (subjective) prior probabilities represent the uncertainty in your knowledge about the true value of θ. It is your knowledge about the parameter that is modeled as random, not the parameter itself. The use of priors distinguishes Bayesian statistics from classical statistics. The so-called likelihood $p(data \mid \theta)$ is the probability of finding the observed data for a given value of θ, and $p(data)$ is obtained by averaging $p(data \mid \theta)$ over the prior probabilities $p(\theta)$. This describes in general terms how Bayesian statistics works. The Bayesian approach is used more and more in practice. Nowadays, Bayesian methods are used widely to address pressing questions in diverse application areas such as astrophysics, actuarial sciences, neurobiology, weather forecasting, spam filtering, forensic text analysis, and criminal justice.

2.5 The concept of random variable

In performing a chance experiment, one is often not interested in the particular outcome that occurs but in a specific numerical value associated with that outcome. Any function that assigns a real number to each outcome in the sample space is called a *random variable*.

The concept of random variable is always a difficult concept for the beginner. A random variable is not a variable in the traditional sense of the word and actually it is a little misleading to call it a variable. Intuitively, a random variable is a function that takes on its value by chance. Formally, a random variable is defined as a function that assigns a numerical value to each element of the sample space. The observed value, or realization, of a random variable is completely determined by the realized outcome of the chance experiment and consequently probabilities can be assigned to the possible values of the random variable. A random variable gets its value only *after* the chance experiment has been done. *Before* the chance experiment is done, you can only speak of the probability that the random variable will take on a particular value. It is common to use uppercase letters such as X, Y, and Z to denote random variables, and lowercase letters x, y, and z to denote their numerical values.

You have *discrete* and *continuous* random variables. A discrete random variable can take on only a finite or a countably infinite number of values, whereas a continuous random variable has a continuum of possible values (think of an interval). The number of goals to be scored in a soccer game is a discrete random variable, but the time until radioactive material will emit a particle is a continuous random variable. In this book the emphasis is on discrete random variables, but Section 3.3 also pays attention to continuous distributions.

Suppose that $a_1, a_2, \ldots, a_r$ are the possible values of a discrete random variable X. The notation $P(X = a_k)$ is used for the probability that the random variable X will take on the value a_k. These probabilities satisfy $\sum_{k=1}^{r} P(X = a_k) = 1$ and constitute the so-called *probability mass function* of the random variable X.

Example 2.10. Two dice are rolled. Let the random variable X be the sum of the points. What is the probability mass function of X?

Solution. As in Example 2.1, it is helpful to think of a blue and a red die. The sample space of the chance experiment consists of the 36 outcomes (i, j) for $i, j = 1, \ldots, 6$, where i is the number rolled by the blue die and j is the number rolled by the red die. Each outcome is equally likely and gets assigned a probability of $\frac{1}{36}$. The random variable X takes on the value $i + j$ when the realized outcome is (i, j). The possible values of X are $2, \ldots, 12$. To find the probability $P(X = k)$, you must know the outcomes (i, j) for which the random variable X takes on the value k. For example, X takes on the value 7 for the outcomes $(1, 6)$, $(2, 5)$, $(3, 4)$, $(4, 3)$, $(5, 2)$, and $(6, 1)$. Each of these six outcomes has probability $\frac{1}{36}$. Thus $P(X = 7) = \frac{6}{36}$. Using the same reasoning, you are asked to verify

$$P(X = j) = \begin{cases} (j - 1)/36 & \text{for } 2 \le j \le 7, \\ (13 - j)/36 & \text{for } 8 \le j \le 12. \end{cases}$$

Problem 2.28. Let the random variable X be the largest number rolled in a roll of two fair dice. What is the probability mass function of X? (answer: $P(X = j) = (2j - 1)/36$ for $j = 1, \ldots, 6$)

Problem 2.29. Each week the value of a particular stock either increases by 5% or decreases by 4% with equal chances, regardless of what happened before. The current value of the stock is \$100. What is the probability mass function of the value of the stock two weeks later? (answer: 0.25, 0.50, 0.25 for \$110.25, \$100.80, \$92.16)

Problem 2.30. Two fair coins are tossed. Then just as many dice are rolled as the number of heads that showed up in the coin tossing. What is the probability mass function of the sum of the scores on the rolled dice? (answer: $\frac{1}{4}, \frac{12}{144}, \ldots, \frac{5}{144}, \ldots, \frac{1}{144}$)

2.6 Expected value and standard deviation

Suppose that $a_1, a_2, \ldots, a_r$ are the possible values of the discrete random variable X. The *expected value* (or *mean* or *average value*) of the random variable X is defined by

$$E(X) = a_1 \times P(X = a_1) + a_2 \times P(X = a_2) + \cdots + a_r \times P(X = a_r).$$

In words, $E(X)$ is a weighted average of the possible values that X can take on, where each value is weighted with the probability that X will take on that particular value.[11] The term 'expected value' can be misleading. This term should not be confused with the term 'most probable value'. Consider the following simple gambling game. A fair die is rolled. If a 6 appears, you get paid \$3; otherwise, you have to pay 60 cents. Define the random variable X as your gain (in dollars). Then X has the possible values 3 and -0.6 with respective probabilities $\frac{1}{6}$ and $\frac{5}{6}$. The most probable value of X is -0.6, but the expected value is

$$E(X) = 3 \times \frac{1}{6} - 0.6 \times \frac{5}{6} = 0.$$

As another example, let the random variable Y be the number of points in a single roll of a fair die. Then,

$$E(Y) = 1 \times \frac{1}{6} + 2 \times \frac{1}{6} + 3 \times \frac{1}{6} + 4 \times \frac{1}{6} + 5 \times \frac{1}{6} + 6 \times \frac{1}{6} = 3.5.$$

Using these two examples, the term 'average value' instead of 'expected value' can be explained. Suppose a fair die is repeatedly rolled. It will come as no surprise that the fraction of the number of rolls with outcome j goes to $\frac{1}{6}$ for all $j = 1, 2, \ldots, 6$ if the number of rolls gets larger and larger. Thus, if the number of rolls gets very large, the average number of points obtained per roll tends to $\frac{1}{6}(1 + 2 + \cdots + 6) = 3.5$, which is the expected value for a single roll. In the gambling game, your average gain per play tends to 0 if

---

[11]The idea of an expected value appears in the 1654 Pascal–Fermat correspondence and this idea was elaborated on by the Dutch astronomer Christiaan Huygens (1625–1695) in his famous 1657 book *Ratiociniis de Ludo Aleae* (On Reasoning in Games of Chance).

the number of plays gets larger and larger. This is known as the *law of large numbers*. This law can be mathematically proved from the axioms and definitions of probability.

In many practical problems, it is helpful to interpret the expected value of a random variable as a long-term average. This is the case in the following example, which has its origin in World War II when a large number of soldiers had their blood tested for syphilis.

Example 2.11. A large number of individuals must undergo a blood test for a certain disease. The probability that a randomly selected person will have the disease is $p = 0.005$. In order to reduce costs, it is decided that the large group should be split into smaller groups, each made up of r persons, after which the blood samples of the r persons will be pooled and tested as one. The pooled blood samples will only test negative (disease free) if all of the individual blood samples were negative. If a test returns a positive result, then all of the r samples from that group will be retested, individually. What is the expected value of the number of tests that will have to be performed on one group of r individuals?

Solution. Define the random variable X as the number of tests that will have to be performed on a group of r individuals. The random variable X has the two possible values 1 and $r + 1$. The probability that X will take on the value 1 is equal to the probability that each individual blood sample will test negative, and this probability is $(1 - p) \times (1 - p) \times \cdots \times (1 - p) = (1 - p)^r$. This means that $P(X = 1) = 0.995^r$ and $P(X = r + 1) = 1 - 0.995^r$. Therefore

$$E(X) = 1 \times 0.995^r + (r + 1) \times (1 - 0.995^r).$$

In other words, by pooling the blood samples of the r individuals, an average of $\frac{1}{r}\left(1 \times 0.995^r + (r + 1) \times (1 - 0.995^r)\right)$ tests per individual will be needed when many groups are tested. The average is minimal for $r = 15$ with 0.1391 as minimum value. Thus the pooling of 15 individual blood samples saves about 86% on the number of tests necessary.

Example 2.12. You play the following game. A fair coin is tossed. If it lands heads, it will be tossed one more time; otherwise, it will be tossed two more times. You win eight dollars if heads does not come up at all, but you must pay one dollar each time heads does turn up. Is this a fair game?

Solution. The set $\{HH, HT, THH, THT, TTH, TTT\}$ is an obvious choice for the sample space, where H stands for heads and T for tails. The random variable X being defined as your net winnings in a game takes on the value 8 for outcome TTT, the value -1 for each of the outcomes HT, THT and TTH, and the value -2 for each of the outcomes HH and THH. The probability $\frac{1}{2} \times \frac{1}{2} = \frac{1}{4}$ is assigned to each of the outcomes HH and HT and the probability $\frac{1}{2} \times \frac{1}{2} \times \frac{1}{2} = \frac{1}{8}$ to each of the other four outcomes. Thus

$$P(X = 8) = \frac{1}{8}, \; P(X = -1) = \frac{1}{4} + \frac{1}{8} + \frac{1}{8}, \; \text{and} \; P(X = -2) = \frac{1}{4} + \frac{1}{8}.$$

This gives

$$E(X) = 8 \times \frac{1}{8} - 1 \times \frac{1}{2} - 2 \times \frac{3}{8} = -\frac{1}{4}.$$

The game is not fair. In the long run, you lose a quarter of a dollar per game.

Linearity property and the substitution formula

A very useful property of the expected value is the *linearity property*. This property states that

$$\boxed{E(aX + bY) = aE(X) + bE(Y)}$$

for any two random variables X and Y and any constants a and b, regardless whether or not there is dependence between X and Y. This result is valid for any type of random variables, as long as the expected values $E(X)$ and $E(Y)$ are well-defined. For the special case of discrete random variables each having only a finite number of possible values, a proof will be given in the appendix of this chapter. More generally, for any random variables $X_1, \ldots, X_n$ and constants $c_1, \ldots, c_n$,

$$\boxed{E(c_1 X_1 + \cdots + c_n X_n) = c_1 E(X_1) + \cdots + c_n E(X_n).}$$

Another important formula is the *substitution formula*, which will also be proved in the appendix. Suppose that $g(x)$ is a given function, say $g(x) = x^2$ or $g(x) = \sin(x)$. Then $g(X)$ is also a random variable when X is a random variable. If X is a discrete random variable with possible values $a_1, \ldots, a_r$, then the substitution formula tells you that the expected value of $g(X)$ can be calculated through the intuitive formula (often called the *law of the unconscious statistician*):

$$E\big[g(X)\big] = \sum_{j=1}^{r} g(a_j) P(X = a_j).$$

Thus you do not need to know the probability mass function of the random variable $g(X)$ in order to calculate its expected value. This is a very useful result. A special case of the substitution formula is

$$E(aX + b) = aE(X) + b \text{ for any constants } a \text{ and } b.$$

In many cases the expected value of a random variable can be calculated by writing the random variable as a sum of *indicator random variables* that can take on only the values 0 and 1. This powerful trick is illustrated in the next example dealing with the celebrated *coupon collector's problem*.

Example 2.13. Each box of a newly launched breakfast cereal contains one baseball card, which is equally likely to be one of 50 different cards. What is the expected number of cards you are still missing after having purchased 100 boxes?

Solution. Define the random variable X as the number of cards you are still missing after having purchased 100 boxes. It is quite complicated to get $E(X)$ through the probability mass function of X. However, it is very easy to calculate $E(X)$ through the trick of writing X as $X = I_1 + I_2 + \cdots + I_{50}$, where $I_k = 1$ if the kth card is missing and $I_k = 0$ otherwise. By the linearity property of the expected value,

$$E(X) = \sum_{k=1}^{50} E(I_k).$$

The $E(I_k)$ are easy to get. For any k, the probability that a box does not contain the kth card is $\frac{49}{50}$ and so $P(I_k = 1) = \left(\frac{49}{50}\right)^{100}$. Since

$$E(I_k) = 1 \times P(I_k = 1) + 0 \times P(I_k = 0) = P(I_k = 1),$$

you have $E(I_k) = \left(\frac{49}{50}\right)^{100}$ for all k. Therefore the expected value of the number of missing cards is $\sum_{k=1}^{50} E(I_k) = 50 \times \left(\frac{49}{50}\right)^{100} = 6.631$.

The coupon collector's problem from Example 2.13 is an application of the *balls-and-bins model*: each of $b = 100$ balls (purchases) is put randomly into one of $n = 50$ bins (baseball cards).

Variance and standard deviation

The expected value is an informative statistic in chance experiments that can be repeated indefinitely often. In other situations, it may dangerous to rely merely on the expected value. Think of non-swimmer going into a lake that is, on average, 30 centimeters deep. A measure for the variability of a random variable X around its expected value $\mu = E(X)$ is the *variance* that is defined by

$$\boxed{\operatorname{var}(X) = E\left[(X - \mu)^2\right].}$$

To illustrate, consider again random variable X with $P(X = 3) = \frac{1}{6}$ and $P(X = -0.6) = \frac{5}{6}$. The random variable X has the expected value $\mu = 0$ and its variance is equal to

$$E[(X - \mu)^2] = (3 - 0)^2 \times \frac{1}{6} + (-0.6 - 0)^2 \times \frac{5}{6} = 1.8.$$

You might wonder why not $E(|X - \mu|)$ is used as measure for the spread of X. One answer is that this measure has not such nice mathematical properties as the mean square of the deviations, but the use of the variance over the mean absolute error is really justified by the normal probability distribution and the central limit theorem, which will be discussed in Section 3.3.

The notation $\sigma^2(X)$ is often used for $\operatorname{var}(X)$. Alternatively, $\operatorname{var}(X)$ can be calculated as

$$\boxed{\operatorname{var}(X) = E(X^2) - \mu^2,}$$

as can be seen by writing $(X - \mu)^2$ as $X^2 - 2\mu X + \mu^2$ and using the linearity property of expectation. Also, using this property and the definition of variance, you can readily verify that

$$\boxed{\text{var}(aX + b) = a^2\text{var}(X) \text{ for any constants } a \text{ and } b.}$$

By the substitution formula, the variance of a discrete random variable X with $a_1, a_2, \ldots, a_r$ as possible values can be calculated as

$$\text{var}(X) = \sum_{j=1}^{r} (a_j - \mu)^2 P(X = a_j) \text{ or } \text{var}(X) = \sum_{j=1}^{r} a_j^2 P(X = a_j) - \mu^2.$$

The variance of a random variable X has not the same units (e.g., dollar or cm) as $E(X)$. For example, if X is expressed in dollars, then $\text{var}(X)$ has (dollars)^2 as dimension. Therefore one often uses the *standard deviation* that is defined as the square root of the variance. The standard deviation of X is denoted by $\sigma(X)$:

$$\boxed{\sigma(X) = \sqrt{\text{var}(X)}.}$$

As an illustration, suppose that X is the number of points to be obtained in a single roll of a die. Since $E(X) = 3.5$, the variance and standard deviation of X are $\sum_{j=1}^{6}(j - 3.5)^2 \frac{1}{6} = \frac{35}{12}$ and $\sqrt{\frac{35}{12}}$.

Example 2.14. Joe and his friend bet every week whether the Dow Jones index will have risen at the end of the week or not. Both put $10 in the pot. Joe observes that his friend is just guessing and is making his choice by the toss of a fair coin. Joe asks his friend if he could contribute $20 to the pot and submit his guess together with that of his brother. The friend agrees. In each week, however, Joe's brother submits a prediction opposite to that of Joe. If there is only one correct prediction, the entire pot goes to that prediction. If there is more than one correct prediction, the pot is split evenly between the correct predictions. How favorable is the game to the brothers?

Solution. Let the random variable X denote the payoff to Joe and his brother in any given week. Either Joe or his brother will have a correct prediction. They win the $30 pot if Joe's friend is wrong; otherwise, they share the pot with Joe's friend. The possible values of X are 30 and 15 dollars. Each of these two values is equally likely, since Joe's friend makes his prediction by the toss of a coin. Thus

$$E(X) = 30 \times \frac{1}{2} + 15 \times \frac{1}{2} = 22.5 \text{ dollars.}$$

Joe and his brother have an expected profit of $E(X - 20) = 2.5$ dollars, each week. To obtain $\sigma(X)$, you first calculate

$$E(X^2) = 900 \times \frac{1}{2} + 225 \times \frac{1}{2} = 562.5 \text{ (dollars)}^2.$$

This gives $\text{var}(X) = 562.5 - 22.5^2 = 56.25 \text{ (dollars)}^2$. Thus

$$\sigma(X) = \sqrt{56.25} = 7.5 \text{ dollars.}$$

Since $\sigma(X - 20) = \sigma(X)$, the standard deviation of the profit $X - 20$ for Joe and his brother is also equal to 7.5 dollars.

Problem 2.31. You roll a die. If a 6 six appears, you win $10. If not, you roll the die again. If a 6 appears the second time, you win $5. If not, you win nothing. What are the expected value and the standard deviation of your winnings? (answer: $2.36 and $3.82)

Problem 2.32. Consider again Problem 2.28. What are the expected value and standard deviation of X? (answer: 4.472 and 1.404)

Problem 2.33. Consider again Problem 2.30. What are the expected value and the standard deviation of the sum of the scores on the rolled dice? (answer: 3.5 and 3.001)

Problem 2.34. Investment A has a 0.8 probability of a $2 000 profit and a 0.2 probability of a $3 000 loss. Investment B has a 0.2 probability of a $5 000 profit and a 0.8 probability of a zero profit. Verify that both investments have the same expected value and the same standard deviation for the net profit. (answer: expected value is $1 000 and standard deviation is $2 000)

Problem 2.35. There are four courses having 15, 20, 75, and 125 students. No student takes more than one course. Let the random variable X be the number of students in a randomly chosen class and Y be the number of students in the class of a randomly chosen student. Can you explain beforehand why $E(Y)$ is larger than $E(X)$? What are $E(X)$ and $E(Y)$? (answer: 58.750 and 93.085) What are the coefficients of variation of X and Y?[12] (answer: 0.764 and 0.415)

Problem 2.36. You throw darts at a circular target on which two concentric circles of radius 1 cm and 3 cm are drawn. The target itself has a radius of 5 cm. You receive 15 points for hitting the target inside the smaller circle, 8 points for hitting the middle annular region, and 5 points for hitting the outer annular region. The probability of hitting the target at all is 0.75. If the dart hits the target, then the hitting point is a random point on the target. Let the random variable X be the number of points scored on a single throw of the dart. What is the expected value of X? (answer: 4.77)

Problem 2.37. Shuffle an ordinary deck of 52 playing cards containing four aces. Then turn up the cards from the top until the first ace appears. What is the expected number of cards that need to be turned over? (answer: 10.6)

Problem 2.38. A bowl has 10 white and 2 red balls. You pick m balls at random, where m can be chosen at your discretion. If each ball picked is white, you win $\$m$; otherwise, you win nothing. What value of m maximizes your expected winnings? (answer: $m = 4$)

Problem 2.39. You have five distinct pairs of socks in a drawer. The socks are not folded in pairs. You pick socks out of the drawer, one at a time and at random. What are the expected value and the standard deviation of the number of socks you must pick out of the drawer in order to get a matching pair? (answer: 4.06 and 1.19)

Problem 2.40. In a barn, 100 chicks sit peacefully in a circle. Suddenly, each chick randomly pecks the chick immediately to its

---

[12]The coefficient of variation of a positive random variable V is defined as $\frac{\sigma(V)}{E(V)}$.

left or right. What is the expected value of the number of unpecked chicks? (answer: 25)

Problem 2.41. What is the expected number of different values that come up when six fair dice are rolled? (answer: 3.991)

Problem 2.42. In the lotto 6/42 a player chooses six different numbers from $1, \ldots, 42$. Suppose players have filled in 5 million tickets with random picks. What is the expected value of the number of different six-number combinations filled in? (answer: $3\,223\,398$).

Problem 2.43. What is the expected value of the largest number drawn in a lotto 6/49 draw? (answer: 42.857)

Problem 2.44. Let the random variable X have the so-called *Bernoulli distribution* (named after the Swiss mathematician Jakob Bernoulli (1654–1705), the originator of the law of large numbers):

$$P(X = 1) = p \text{ and } P(X = 0) = 1 - p.$$

Verify that $E(X) = p$ and $\sigma(X) = \sqrt{p(1-p)}$.

Problem 2.45. A random variable X with the probability mass function

$$P(X = k) = (1 - p)^{k-1}p \quad \text{for } k = 1, 2, \ldots$$

is said to have a *geometric probability distribution*.[13] Use results for the geometric series in Section 1.2 to verify that $E(X) = \frac{1}{p}$ and $\sigma(X) = \frac{1}{p}\sqrt{1-p}$.

Problem 2.46. What is the expected number of boxes that must be purchased in order to get a complete set of cards in the coupon collector's problem from Example 2.13? (answer: 224.96)

Problem 2.47. What is the expected number of rolls of a fair die it takes to see all six sides of the die at least once? (answer: 14.7)

---

[13]This is the probability distribution of the number of trials until the first success occurs in a sequence of independent trials each having the same success probability p. A very useful probability model!

Problem 2.48. You play a game in which you can pick a random number from 1 to 25 as often as you wish. Each pick costs you one dollar. If you decide to stop, you get paid in dollars your last picked number. What strategy maximizes your expected net payoff? (answer: stop if your picked number is ≥ 19)

Problem 2.49. On a game show, you can bet on one of the numbers 1 to 100. Then, a random number is generated from 1 to 100. If your guess is less than the randomly chosen number, you win in dollars the square of your guess; otherwise, you win nothing. What number should you guess to maximize your expected winning? (answer: 67)

2.7 Independent random variables and the square root law

As you have seen in Section 2.6, it is always true that $E(X + Y) = E(X) + E(Y)$ for any two random variables X and Y. A similar result for the variance is in general not true. You can see this from the example with $P(X = 1) = P(X = -1) = \frac{1}{2}$ and $Y = -X$. In this example, $\text{var}(X + Y) = 0$ and $\text{var}(X) = \text{var}(Y) = 1$ (verify!). The reason that $\text{var}(X + Y)$ is not equal to $\text{var}(X) + \text{var}(Y)$ is that X and Y are not independent of each other. Two random variables X and Y are said to be *independent* of each other if

$$P(X \leq x \text{ and } Y \leq y) = P(X \leq x)P(Y \leq y) \quad \text{for all } x \text{ and } y \,.$$

For discrete random variables X and Y, an alternative definition of independence is $P(X = x \text{ and } Y = y) = P(X = x)P(Y = y)$ for all x and y. Then the following result holds

$$\text{var}(aX + bY) = a^2\text{var}(X) + b^2\text{var}(Y) \quad \text{for independent } X \text{ and } Y,$$

where a and b are constants. This result is true for any type of random variables. For the special case of discrete random variables X and Y, a proof is given in the appendix to this chapter. More generally, for independent $X_1, \ldots, X_n$ and constants $c_1, \ldots, c_n,$ [14]

$$\text{var}(c_1 X_1 + \cdots + c_n X_n) = c_1^2 \, \text{var}(X_1) + \cdots + c_n^2 \, \text{var}(X_n).$$

---

[14]In general, random variables $X_1, \ldots, X_n$ are said to be independent if $P(X_1 \leq x_1$ and $\ldots$ and $X_n \leq x_n) = P(X_1 \leq x_1) \cdots P(X_n \leq x_n)$ for all $x_1, x_2, \ldots, x_n$.

As an illustration, what is the standard deviation of the sum of a roll of two dice? Let X be the number rolled on the first die and Y be the number rolled on the second die. The sum of a roll of two dice is given by $X + Y$. As calculated before, $\text{var}(X) = \text{var}(Y) = \frac{35}{12}$. The random variables X and Y are independent of each other. Then $\text{var}(X + Y) = \text{var}(X) + \text{var}(Y) = \frac{35}{6}$. Thus the standard deviation of the sum of a roll of two dice is $\sigma(X + Y) = \sqrt{\frac{35}{6}} \approx 2.415$ points.

The square root law for the standard deviation

Let $X_1, \ldots, X_n$ be independent random variables each having standard deviation σ. Then, by $\text{var}\left(\sum_{k=1}^{n} X_k\right) = \sum_{k=1}^{n} \text{var}(X_k) = n\sigma^2$ and $\text{var}\left(\frac{1}{n}\sum_{k=1}^{n} X_k\right) = \frac{1}{n^2}\sum_{k=1}^{n} \text{var}(X_k) = \frac{\sigma^2}{n}$, you find

$$\sigma\left(\sum_{k=1}^{n} X_k\right) = \sigma\sqrt{n} \quad \text{and} \quad \sigma\left(\frac{1}{n}\sum_{k=1}^{n} X_k\right) = \frac{\sigma}{\sqrt{n}}.$$

This is called the *square root law* (or the $\sqrt{n}$-law) for the standard deviation. It is an extremely important result in probability and statistics. In Figure 2 an experimental demonstration of the $\sqrt{n}$-law is given. For each of the values $n = 1$, 4, 16 and 64, one hundred random outcomes of the so-called sample mean $\frac{1}{n}\sum_{k=1}^{n} X_k$ are given for the case that the X_k have a same probability distribution with expected value 0 and standard deviation 1 (the chosen distribution is the standard normal distribution, which will be discussed in Section 3.3). The observed outcomes are determined by computer simulation. You see from the figure that the bandwidths within which the simulated outcomes lie are indeed reduced by a factor of about 2 when the sample sizes increase by a factor of 4.

The $\sqrt{n}$-law is sometimes called the De Moivre's equation, after Abraham de Moivre (1667–1754).[15] This formula had an immediate

---

[15]The French-born Abraham de Moivre was the leading probabilist of the eighteenth century and lived most of his life in England. The protestant De Moivre left France in 1688 to escape religious persecution. He was a good friend of Isaac Newton and supported himself by calculating odds for gamblers and insurers and by giving private lessons to students.

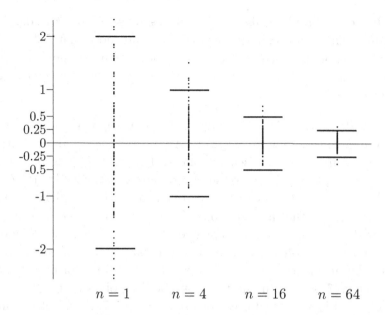

$n = 1 \qquad n = 4 \qquad n = 16 \qquad n = 64$

Figure 2: An illustration of the square root law

impact on methods used to inspect gold coins struck at the London Mint. The standard gold weight, per coin, was 128 grains (this was equal to 0.0648 gram), and the allowable deviation from this standard was $\frac{1}{400}$ of that amount, or 0.32 grains. A test case of 100 coins was periodically performed on coins struck, their total weight then being compared with the standard weight of 100 coins. The gold used in the striking of coins was the property of the king, who sent inspectors to discourage minting mischief. The royal watch dogs had traditionally allowed a deviation of $100 \times 0.32 = 32$ grains in the weight of 100 inspected coins. Directly after De Moivre's publication of the square root formula, the allowable deviation in the weight of 100 coins was changed to $\sqrt{100} \times 0.32 = 3.2$ grains; alas for the English monarchy, previous ignorance of the square root formula had cost them a fortune in gold.

The square root law has many applications, providing explanation, for example, for why city or hospital size is important for measuring crime statistics or death rates after surgery. Small hospitals, for example, are more likely than large ones to appear at the top or bottom of ranking lists. This makes sense if you consider that, when tossing a fair coin, the probability that more than 70%, or less than 30%, of the tosses will turn up heads is much larger for 10 coin tosses, than for 100. The smaller the numbers, the larger the chance fluctuations!

Example 2.15. 'Unders and Overs' is a popular game formerly played during open house events at American schools, for the purpose of adding money to the school coffers. The game is played with two dice and a playing board divided into three sections: 'Under 7', '7', and 'Over 7'. The two dice are rolled, and players place chips on one or more of the three sections. Chips may be placed on the game board for 1 dollar apiece. For every chip placed in the 'Under 7' section, the payoff is 2 dollars if the total number of points rolled with the dice is less than 7. The payoff is the same for every chip in the 'Over 7' section if the total number of points is higher than 7. The payoff is 5 dollars for each chip placed on '7' if the total number of points is 7. A popular strategy is to place 1 chip on each of the three sections. Suppose that 500 rounds of the game are played, using this strategy. In each round there is a single player. What are the expected value and standard deviation of the net amount taken in by the school as a result of the 500 bets?

Solution. Let the random variable X be the net profit of the school in a single play of the game. The random variable X can take on the two values \$1 and $-\$2$. The net profit is $3 - 5 = -2$ if the sum of the dice is 7 and is $3 - 2 = 1$ otherwise. Using the results of Example 2.10, you find $P(X = -2) = \frac{6}{36}$ and $P(X = 1) = 1 - \frac{6}{36} = \frac{30}{36}$. Thus

$$E(X) = -2 \times \frac{6}{36} + 1 \times \frac{30}{36} = \frac{1}{2}.$$

Let's get $\sigma(X)$ from the formula $\sigma^2(X) = E(X^2) - \big(E(X)\big)^2$. Since $E(X^2) = 4 \times \frac{6}{36} + 1 \times \frac{30}{36} = \frac{3}{2}$, you get $\sigma^2(X) = \frac{3}{2} - (\frac{1}{2})^2 = \frac{5}{4}$ and so

$$\sigma(X) = \frac{1}{2}\sqrt{5}.$$

The total net profit of the school is $X_1 + \cdots + X_{500}$ dollars, where X_i is the net profit of the school in the ith round. The random variables

$X_1, \ldots, X_{500}$ are each distributed as X and are independent of each other. Using the linearity property of expectation,

$$E(X_1 + \cdots + X_{500}) = 500 \times \frac{1}{2} = 250 \text{ dollars.}$$

By the $\sqrt{n}$-law for the standard deviation,

$$\sigma(X_1 + \cdots + X_{500}) = \frac{1}{2}\sqrt{5} \times \sqrt{500} = 25 \text{ dollars.}$$

Problem 2.50. A fair die will be thrown 100 times. What are the expected value and the standard deviation of the average number of points per throw? (answer: 3.5 and 0.171)

Problem 2.51. The Mang Kung dice game is played with six non-traditional dice. Each of the six dice has five blank faces and one face marked with one of the numbers 1 up to 6 such that no two dice have the same number. What are the expected value and the standard deviation of the total number of points showing up when the six dice are rolled? (answer: 3.5 and 3.555).

Problem 2.52. Consider Problem 2.46 again. Use results from Problem 2.45 to calculate the standard deviation of the number of purchases needed to get a complete set of cards. (answer: 61.951)

Problem 2.53. Consider Problem 2.47 again. What is standard deviation of the number of rolls needed to get all six possible outcomes? (answer: 6.244)

2.8 Generating functions[16]

Counting was used in Example 2.10 to calculate the probability mass function of the sum of the upturned faces when rolling two dice. The counting approach becomes very tedious when asking about three dice, not to mention five or ten dice. A kind of magic approach to handle these cases is the generating function approach. Generating

---

[16]This section can be skipped at first reading.

functions were introduced by the Swiss genius Leonhard Euler (1707–1783) to facilitate calculations in counting problems. The idea is very simple and illustrates that simple ideas are often the best.

The generating function $G_X(z)$ of a non-negative, integer-valued random variable X is defined by the power series

$$G_X(z) = \sum_{k=0}^{\infty} P(X = k) z^k \quad \text{for } |z| \leq 1.$$

The generating function $G_X(z)$ uniquely determines the probability mass function of X: taking the rth derivative of $G_X(z)$ at $z = 0$ gives $r!P(X = r)$ (verify). Also, $G'_X(1) = E(X)$. If X and Y are independent random variables on the non-negative integers, then

$$G_{X+Y}(z) = G_X(z)G_Y(z) \quad \text{for } |z| \leq 1,$$

that is, the generating function of the sum $X + Y$ is the product of the generating functions of X and Y. This is a very important result. To prove the result, it is first noted that, by the substitution rule from Section 2.6,

$$E(z^X) = \sum_{k=0}^{\infty} z^k P(X = k) \quad \text{for any } |z| \leq 1,$$

and so $G_X(z) = E(z^X)$ for any $|z| \leq 1$. Moreover, if X and Y are independent random variables, then for any functions f and g the random variables $f(X)$ and $g(Y)$ are independent. Taking this fact for granted, it follows that the random variables z^X and z^Y are independent for any $|z| \leq 1$. In the appendix to this chapter it will be proved that $E(VW) = E(V)E(W)$ for independent random variables V and W. Thus, for any $|z| \leq 1$,

$$G_{X+Y}(z) = E(z^{X+Y}) = E(z^X z^Y) = E(z^X)E(z^Y) = G_X(z)G_Y(z).$$

Let's now apply the generating function method to the calculation of the probability mass function of the sum of the upturned faces when rolling three dice. Let X_i be the upturned face value of the ith die for $i = 1, 2, 3$. The generating function of X_i is given by

$$G_{X_i}(z) = \frac{1}{6}z + \frac{1}{6}z^2 + \frac{1}{6}z^3 + \frac{1}{6}z^4 + \frac{1}{6}z^5 + \frac{1}{6}z^6 \quad \text{for } i = 1, 2, 3.$$

Since X_1, X_2 and X_3 are independent, the generating function of the sum of the upturned face values is given by

$$G_{X_1+X_2+X_3}(z) = \left(\frac{1}{6}z + \frac{1}{6}z^2 + \frac{1}{6}z^3 + \frac{1}{6}z^4 + \frac{1}{6}z^5 + \frac{1}{6}z^6\right)^3 \quad \text{for } |z| \le 1.$$

Using standard software to calculate product polynomials,

$$\begin{aligned}
G_{X_1+X_2+X_3}(z) = {} & \frac{1}{216}z^3 + \frac{1}{72}z^4 + \frac{1}{36}z^5 + \frac{5}{108}z^6 + \frac{5}{72}z^7 + \frac{7}{72}z^8 \\
& + \frac{25}{216}z^9 + \frac{1}{8}z^{10} + \frac{1}{8}z^{11} + \frac{25}{216}z^{12} + \frac{7}{72}z^{13} \\
& + \frac{5}{72}z^{14} + \frac{5}{108}z^{15} + \frac{1}{36}z^{16} + \frac{1}{72}z^{17} + \frac{1}{216}z^{18}.
\end{aligned}$$

The probability mass function of the sum can be directly read off from this expansion. The coefficient of z^k gives the probability that the sum $X_1 + X_2 + X_3$ takes on the value k for $k = 3, \dots, 18$.

Problem 2.54. You have 8 symmetric six-sided dice. Five of these dice have the number six on two of the faces and the other three have the number six on three of the faces. What is the probability mass function of the number of sixes appearing when rolling the 8 dice? (answer: $\left(\frac{4}{243}, \frac{22}{243}, \frac{52}{243}, \frac{23}{81}, \frac{25}{108}, \frac{77}{648}, \frac{73}{1944}, \frac{13}{1944}, \frac{1}{1944}\right)$)

Problem 2.55. A Poisson distributed random variable X with parameter λ has probability mass function $P(X = k) = e^{-\lambda}\lambda^k/k!$ for $k = 0, 1, \dots$. Verify that $G_X(z) = e^{-\lambda(1-z)}$ is the generating function of X. What is the probability mass function of the sum of two independent Poisson random variables X and Y with parameters α and β? (answer: Poisson distribution with parameter $\alpha + \beta$)

Problem 2.56. Let $X_1, \dots, X_5$ be independent Bernoulli variables with $P(X_i = 1) = \frac{1}{i+1}$ and $P(X_i = 0) = 1 - \frac{1}{i+1}$ for $i = 1, \dots, 5$. What is the probability mass function of the sum $X_1 + \cdots + X_5$? (answer: $\left(\frac{1}{6}, \frac{137}{360}, \frac{5}{16}, \frac{17}{144}, \frac{1}{48}, \frac{1}{720}\right)$)

Appendix: Proofs for expected value and variance

This appendix proves the properties that were given in the Sections 2.6 and 2.7 for the expected value and variance and presents some additional results. The proofs will be given for the case that X and Y are discrete random variables that can take on only a finite number of values. Let I be the set of possible values of X and J be the set of possible values of Y. For the moment, you are asked to take for granted the following basic result that will be proved below: for any function $g(x, y)$, the expected value of the random variable $g(X, Y)$ is given by

$$E\big(g(X,Y)\big) = \sum_{x \in I} \sum_{y \in J} g(x,y) P(X = x \text{ and } Y = y).$$

A double sum $\sum_{i=1}^{n} \sum_{j=1}^{m} a_{ij}$ should be read as $\sum_{i=1}^{n}(a_{i1} + \cdots + a_{im})$. You can always interchange the order of summation when there are finitely many terms: $\sum_{i=1}^{n} \sum_{j=1}^{m} a_{ij} = \sum_{j=1}^{m} \sum_{i=1}^{n} a_{ij}$.

The following properties can now be easily proved:

Property 1. $E(aX + bY) = aE(X) + bE(Y)$ for constants a and b.

Property 2. $E(XY) = E(X)E(Y)$ if X and Y are independent of each other.

Property 3. $\text{var}(aX + bY) = a^2\text{var}(X) + b^2\text{var}(Y)$ for any constants a and b if X and Y are independent of each other.

Proof of Property 1. Using the basic result for $E\big(g(X,Y)\big)$ for $g(x, y) = ax + by$, you get that $E(aX + bY)$ is equal to

$$\sum_{x \in I} \sum_{y \in J} (ax + by) P(X = x \text{ and } Y = y)$$

$$= \sum_{x \in I} \sum_{y \in J} ax\, P(X = x \text{ and } Y = y) + \sum_{x \in I} \sum_{y \in J} by\, P(X = x \text{ and } Y = y)$$

$$= a \sum_{x \in I} x \sum_{y \in J} P(X = x \text{ and } Y = y) + b \sum_{y \in J} y \sum_{x \in I} P(X = x \text{ and } Y = y),$$

where the order of summation is interchanged in the second term of the last equation. Next you use the formula

$$P(X = x) = \sum_{y \in J} P(X = x \text{ and } Y = y).$$

This formula is a direct consequence of Axiom 3 in Section 2.1. A similar formula applies to $P(Y = y)$. This gives

$$E(aX+bY) = a \sum_{x \in I} xP(X = x) + b \sum_{y \in J} yP(Y = y) = aE(X) + bE(Y).$$

Proof of Property 2. The definition of independent random variables X and Y is $P(X = x \text{ and } Y = y) = P(X = x)P(Y = y)$ for all possible values x and y. Next, by applying the basic result for $E(g(X,Y))$ with $g(x, y) = xy$, you find that $E(XY)$ is equal to

$$\sum_{x \in I} \sum_{y \in J} xyP(X = x \text{ and } Y = y) = \sum_{x \in I} \sum_{y \in J} xyP(X = x)P(Y = y)$$

$$= \sum_{x \in I} xP(X = x) \sum_{y \in J} yP(Y = y) = E(X)E(Y).$$

Proof of Property 3. For ease of notation, write $E(X)$ as μ_X and $E(Y)$ as μ_Y. Using the alternative definition $\text{var}(V) = E(V^2) - \mu^2$ for the variance of a random variable V with expected value μ and using Property 1, you get

$$\text{var}(aX + bY) = E((aX + bY)^2) - (E(aX + bY))^2$$
$$= a^2 E(X^2) + 2abE(XY) + b^2 E(Y^2) - (a\mu_X + b\mu_Y)^2.$$

Next you use the independence of X and Y. This gives $E(XY) = E(X)E(Y)$, by Property 2, and so $\text{var}(aX + bY)$ is equal to

$$a^2 E(X^2) + 2ab\mu_X\mu_Y + b^2 E(Y^2) - a^2 \mu_X^2 - 2ab\mu_X\mu_Y - b^2 \mu_Y^2$$
$$= a^2[E(X^2) - \mu_X^2] + b^2[E(Y^2) - \mu_Y^2] = a^2 \text{var}(X) + b^2 \text{var}(Y).$$

A special case of Property 3 is that $\text{var}(aX + b) = a^2\text{var}(X)$ for any constants a and b. This follows by taking for Y a degenerate random variable with $P(Y = 1) = 1$ for which $\text{var}(Y) = 0$ (verify).

It remains to prove the formula $\sum_{(x,y)} g(x,y)P(X = x \text{ and } Y = y)$ for $E(g(X,Y))$. The trick is to define the random variable $Z = g(X,Y)$. Then, $\sum_{(x,y)} g(x,y)P(X = x \text{ and } Y = y)$ can be written as

$$\sum_z \Big[\sum_{(x,y):g(x,y)=z} g(x,y)P(X = x \text{ and } Y = y)\Big]$$

$$= \sum_z z \sum_{(x,y):g(x,y)=z} P(X = x \text{ and } Y = y) = \sum_z zP(Z = z)$$

$$= E(Z) = E\big(g(X,Y)\big).$$

Taking $Y = X$ and $g(x,y) = g(x)$, this result proves the substitution formula in Section 2.6 as special case.

As an encore, we give two famous inequalities involving the concepts of the expected values and variance. For a *non-negative* random variable Y with finite expected value, *Markov's inequality* states that

$$\boxed{P(Y \geq a) \leq \frac{E(Y)}{a} \quad \text{for any constant } a > 0.}$$

The proof is surprisingly simple. For fixed $a > 0$, let the indicator variable I be equal to 1 if $Y \geq a$ and 0 otherwise. Then $E(I) = P(Y \geq a)$. By the non-negativity of Y, you have that $Y \geq aI$, and so

$$E(Y) \geq aE(I) = aP(Y \geq a),$$

which gives the inequality. Beauty in simplicity! The inequality was proved by the Russian mathematician Andrey A. Markov (1856–1922) in 1889 and it is the basis for many other useful inequalities in probability. For any random variable X having finite expected value μ and finite variance σ^2, *Chebyshev's inequality* states that

$$\boxed{P(|X - \mu| \geq c) \leq \frac{\sigma^2}{c^2} \quad \text{for any constant } c > 0.}$$

This inequality is named after the Russian mathematician Pafnuty L. Chebyshev (1821–1894) who proved the inequality in 1867. A simpler proof was later given by his famous student Andrey Markov: taking $Y = (X - \mu)^2$ and $a = c^2$ in Markov's inequality, you get

$$P(|X - \mu| \geq c) = P\big((X - \mu)^2 \geq c^2\big) \leq \frac{E[(X - \mu)^2]}{c^2} = \frac{\sigma^2}{c^2}.$$

Covariance and correlation coefficient

How to calculate $\text{var}(X + Y)$ if X and Y are not independent random variables? To do this, you need the concept of covariance. The *covariance* of two random variables X and Y is defined by

$$\text{cov}(X, Y) = E\big[(X - E(X))(Y - E(Y))\big].$$

Note that $\text{cov}(X, X) = \text{var}(X)$. The formula for $\text{cov}(X, Y)$ can be written in the equivalent form

$$\text{cov}(X, Y) = E(XY) - E(X)E(Y),$$

by writing $(X - E(X))(Y - E(Y))$ as $XY - XE(Y) - YE(X) + E(X)E(Y)$ and using the linearity property of the expectation operator. It is immediate from the definition of covariance that

$$\text{cov}(aX, bY) = ab\,\text{cov}(X, Y)$$

for any constants a and b. Also, using the fact that $E(XY) = E(X)E(Y)$ for independent X and Y (see Property 2), it follows that

$$\text{cov}(X, Y) = 0 \quad \text{if } X \text{ and } Y \text{ are independent random variables.}$$

However, the converse of this result is not always true. As a counterexample, let X take on the values -1 and 1 each with probability $\frac{1}{2}$, and let $Y = X^2$. Then, by $E(X) = 0$ and $E(X^3) = 0$, you have $\text{cov}(X, Y) = E(X^3) - E(X)E(X^2) = 0$, but X and Y are not independent.

The following formula for the variance of any two random variables X and Y can now be formulated:

$$\text{var}(X + Y) = \text{var}(X) + 2\text{cov}(X, Y) + \text{var}(Y).$$

The proof of this result is a matter of some algebra. Put for abbreviation $\mu_X = E(X)$ and $\mu_Y = E(Y)$. Using the definition of variance and the linearity of the expectation operator, you have that

$$\begin{aligned}
\text{var}(X + Y) &= E[(X + Y - (\mu_X + \mu_Y))^2] \\
&= E[(X - \mu_X)^2] + 2E[(X - \mu_X)(Y - \mu_Y)] + E[(Y - \mu_Y)^2] \\
&= \text{var}(X) + 2\text{cov}(X, Y) + \text{var}(Y),
\end{aligned}$$

as was to be verified. The formula for $\text{var}(X+Y)$ can be extended to the case of finitely many random variables:

$$\text{var}(X_1 + \cdots + X_n) = \sum_{i=1}^{n} \text{var}(X_i) + 2 \sum_{i=1}^{n-1} \sum_{j=i+1}^{n} \text{cov}(X_i, X_j).$$

To illustrate, what is the variance of the number of integers that keep their original positions in a random permutation of the integers 1 to n? This quantity can be written as $\sum_{i=1}^{n} X_i$, where $X_i = 1$ if integer i keeps its original position and $X_i = 0$ otherwise. By $E(X_i^2) = E(X_i) = P(X_i = 1) = \frac{1}{n}$ for all i and $E(X_i X_j) = P(X_i = 1$ and $X_j = 1) = \frac{1}{n(n-1)}$ for all $i \neq j$, you get that $\text{var}(X_i) = \frac{1}{n} - \frac{1}{n^2}$ for all i and $\text{cov}(X_i, X_j) = \frac{1}{n(n-1)} - \frac{1}{n^2}$ for all $i \neq j$. This leads to the (surprising) answer $\text{var}\left(\sum_{i=1}^{n} X_i\right) = 1$ (verify!).

Another very important concept in statistics is the concept of correlation coefficient. The units of $\text{cov}(X,Y)$ are not the same as the units of $E(X)$ and $E(Y)$. Therefore it is often more convenient to use the *correlation coefficient* of X and Y. This statistic is defined as

$$\rho(X, Y) = \frac{\text{cov}(X, Y)}{\sigma(X)\sigma(Y)},$$

provided that $\sigma(X) > 0$ and $\sigma(Y) > 0$. The correlation coefficient is a dimensionless quantity with the property that

$$-1 \leq \rho(X, Y) \leq 1.$$

The algebraic proof of this property is omitted. The random variables X and Y are said to be *uncorrelated* if $\rho(X, Y) = 0$. Independent random variables X and Y are always uncorrelated, since their covariance is zero. However, the converse of this result is not always true. Nevertheless, $\rho(X, Y)$ is often used as a measure of the dependence of X and Y.

Problem 2.57. The *least squares regression line* of Y with respect to X is defined by $y = E(Y) + b(x - E(X))$, where the slope b minimizes $E[(Y - E(Y) - b(X - E(X)))^2]$. What is the minimizing value of b? (answer: $b = \rho(X, Y) \frac{\sigma(Y)}{\sigma(X)}$).

Chapter 3

Useful Probability Distributions

This chapter introduces you to three important discrete probability distributions and to three important continuous probability distributions. The discrete distributions are the binomial, hypergeometric and Poisson distributions, while the three continuous distributions are the normal, the uniform and the exponential distributions. Insight is given to the practical relevance of these distributions.

3.1 The binomial and hypergeometric distributions

A random variable X with $0, 1, \ldots, n$ as possible values is said to have a *binomial distribution* with parameters n and p if

$$P(X = k) = \binom{n}{k} p^k (1 - p)^{n-k} \quad \text{for } k = 0, 1, \ldots, n,$$

where $\binom{n}{k}$ is the binomial coefficient. This distribution arises in probability problems that can be formulated within the framework of a sequence of physically independent trials, where each trial has the two possible outcomes 'success' (S) and 'failure' (F). The outcome success occurs with probability p and the outcome failure with probability $1 - p$. The random variable X defined as the total number of successes in n trials has a binomial distribution. This result is easily derived. The probability that a pre-specified sequence of k successes and $n - k$ failures will occur is $p^k (1 - p)^{n-k}$; for example, for $n = 5$ and $k = 3$, the sequence $SSFSF$ will occur with probability $p \times p \times (1 - p) \times p \times (1 - p) = p^3(1 - p)^2$. The total number of possible sequences with k successes and $n - k$ failures is $\binom{n}{k}$, since the binomial coefficient $\binom{n}{k}$ is the total number of ways to choose k different positions from n available positions, see Section 1.1.

The expected value and standard deviation of X are given by

$$\boxed{E(X) = np \quad \text{and} \quad \sigma(X) = \sqrt{np(1 - p)}.}$$

The simplest way to derive these formulas is to use indicator variables. Write X as $X = I_1 + \cdots + I_n$, where I_k is 1 if the kth trial is a success and I_k is 0 otherwise. The Bernoulli variable I_k has $E(I_k) = p$ and $\sigma(I_k) = \sqrt{p(1 - p)}$, see Problem 2.44. Applying the linearity property for expectation and the $\sqrt{n}$-law for the standard deviation (the I_k's are independent of each other), you get

$$E\left(\sum_{k=1}^{n} I_k\right) = \sum_{k=1}^{n} E(I_k) = np \text{ and } \sigma\left(\sum_{k=1}^{n} I_k\right) = \sqrt{p(1 - p)} \times \sqrt{n}.$$

The binomial distribution is a versatile probability distribution and has numerous applications.

Example 3.1. A military early-warning installation is constructed in a desert. The installation consists of seven detectors including two reserve detectors. If fewer than five detectors are working, the installation ceases to function. Every six months an inspection of the installation is mounted and at that time all detectors are replaced by new ones. There is a probability of 0.05 that any given detector will fail between two inspections. The detectors are all in operation and act independently of one another. What is the probability that the system will cease to function between two inspections?

Solution. Let the random variable X denote the number of detectors that will cease to function between two inspections. Then the random variable X has a binomial distribution with parameters $n = 7$ and $p = 0.05$. The probability that the system will cease to function between two inspections is

$$P(X > 2) = \sum_{k=3}^{7} \binom{7}{k} 0.05^k \, 0.95^{7-k} = 0.0038.$$

The binomial distribution can be used to solve the famous *problem of points*. In 1654 this problem was posed to Pascal and Fermat by the compulsory gambler Chevalier de Méré. Mathematics historians believe that the Chevalier posed the following problem: "Two players play a chance game of three points and each player has staked 32 pistoles. How should the sum be divided if they break off prematurely when one player has two points and the other player has one point?" A similar problem was earlier posed by the Italian mathematician Luca Pacioli in 1494 and led to heated discussions among Italian mathematicians in the 16th century, but none of them could come up with a satisfactory answer. The starting insight for Pascal and Fermat was that what is important is not so much the number of points each player has won yet, but the ultimate win probabilities of the players if the game were to continue at the point of stopping. The stakes should be divided in proportion to these win probabilities. Today the solution to the problem is obvious, but it was not at all obvious how to solve the problem in a time that the theory of probability was at an embryonic stage. The next example analyzes the problem of points in a modern outfit.

Example 3.2. In the World Series Baseball, the final two teams play a series consisting of no more than seven games until one of the teams has won four games. The winner takes all of the prize money of $1 000 000. In one such a final, two teams are pitted against one another and the stronger team will win any given game with a probability of 0.55. Unexpectedly, the competition must be suspended when the weaker team leads two games to one. How should the prize money be divided if the remaining games cannot be played?

Solution. At the point of stopping, the weaker team is 2 points away from the required 4 points and the stronger team 3 points. In the actual game at most $2 + 3 - 1 = 4$ more matches would be needed to declare a winner. A trick to solve the problem is to imagine that four additional matches would be played. The probability of the weaker team being the ultimate winner if the original game was to be continued is the same as the probability that the weaker team would win two or more matches in four additional matches (explain!). The latter probability is equal to the binomial probability

$$\sum_{k=2}^{4} \binom{4}{k} 0.45^k \, 0.55^{4-k} = 0.609019.$$

The weaker team should receive \$609 019 and the stronger team \$390 981.

A less famous but still interesting problem from the history of probability is the Newton–Pepys problem. Isaac Newton was not much interested in probability. Nevertheless Newton solved the following dice problem brought to him by Samuel Pepys who was a president of the Royal Society of London and apparently a gambling man. Which game is more likely to win: at least one six in one throw of six dice, at least two sixes in one throw of twelve dice, or at least three sixes in one throw of eighteen dice? What do you think? Pepys believed that the last option was the most favorable one.

The hypergeometric distribution

The hypergeometric distribution is a discrete distribution that is closely related to the binomial distribution. The difference is that the trials in the hypergeometric context are not independent. The hypergeometric distribution describes the probability distribution of the number of successes when sampling *without replacement* from a finite population consisting of elements of two kinds. Think of an urn containing red and white balls or a shipment containing good and defect items. Suppose the population has R elements of the first type (for convenience, called successes) and W elements of the second type (called failures). Let n be the given number of elements that are randomly drawn from the population without replacement. Denote by the random variable X the number of successful elements drawn. Then, X has the *hypergeometric probability distribution*

$$P(X = r) = \frac{\binom{R}{r}\binom{W}{n-r}}{\binom{R+W}{n}} \quad \text{for } r = 0, 1, \ldots, R,$$

with the convention that $\binom{a}{b} = 0$ for $b > a$. The derivation of this distribution was already given in Section 1.1 and went as follows.

Imagine that the elements are labeled so that all elements are distinguishable. Then the total number of possible combinations of n distinguishable elements is $\binom{R+W}{n}$ and among those combinations there are $\binom{R}{r} \times \binom{W}{n-r}$ combinations with r elements of the first type and $n-r$ elements of the second type. Each combination is equally likely and so $P(X = r)$ is the ratio of $\binom{R}{r} \times \binom{W}{n-r}$ and $\binom{R+W}{n}$.

The expected value and the standard deviation of X are

$$E(X) = np \quad \text{and} \quad \sigma(X) = \sqrt{np(1-p)\frac{R+W-n}{R+W-1}},$$

where $p = \frac{R}{R+W}$. The derivation of these formulas requires tedious algebra and is omitted. The hypergeometric distribution can be approximated by a binomial distribution with parameters n and $p = \frac{R}{R+W}$ when n is much smaller than $R + W$.

The best example for the hypergeometric distribution is the lottery, see Section 1.1. The hypergeometric model has many applications and shows up in various disguises, which at first sight have little to do with the classical model of red and white balls in an urn.

Example 3.3. In a close election between two candidates A and B in a small town the winning margin of candidate A is 1 422 to 1 405 votes. However, 101 votes are found to illegal and have to be thrown out. It is not said how the illegal votes are divided between the two candidates. Assuming that the illegal votes are not biased in any particular way and the count is otherwise reliable, what is the probability that the removal of the illegal votes will change the result of the election?

Solution. The problem can be translated into the urn model with 1 422 red and 1 405 white balls. If a is the number of illegal votes for candidate A and b the number of illegal votes for candidate B, then candidate A will have no longer most of the votes only if $a - b \geq 17$. Since $a + b = 101$, the inequality $a - b \geq 17$ boils down to $2a \geq 101 + 17$, or $a \geq 59$. The probability that the removal of the illegal votes will change the election result is the same as the probability of getting 59 or more red balls when randomly picking 101 balls from an urn with 1 422 red and 1 405 white balls. This probability is

$$\sum_{a=59}^{101} \frac{\binom{1\,422}{a}\binom{1\,405}{101-a}}{\binom{2\,827}{101}} = 0.0592.$$

Problem 3.1. A fair coin is to be tossed six times. You win two dollars if heads appears exactly three times (the expected number) and you lose one dollar otherwise. Is this game advantageous to you? (answer: no, your win probability is $\frac{5}{16}$)

Problem 3.2. Each day, the teacher randomly draws the names of four pupils in a class of 25 pupils. The homework of those four pupils is checked. What is the probability that your name will be drawn three or more times in the next five days? (answer: 0.0318)

Problem 3.3. Daily Airlines flies from Amsterdam to London every day. The plane has a passenger capacity of 150. The airline management has made it a policy to sell 157 tickets for this flight in order to protect themselves against no-show passengers. Experience has shown that the probability of a passenger being a no-show is equal to 0.08. The booked passengers act independently of each other. What is the probability that some passengers will have to be bumped from the flight? (answer: 0.0285)

Problem 3.4. Chuck-a-Luck is a carnival game of chance. To play this game, the player chooses one number from the numbers $1, \ldots, 6$. Then three dice are rolled. If the player's number does not come up at all, the player loses 10 dollars. If the chosen number comes up one, two, or three times, the player wins \$10, \$20, or \$30 respectively. What are the expected value and the standard deviation of the win for the house per wager? (answer: \$0.787 and \$11.13)

Problem 3.5. Suppose that 500 debit cards are stolen in a certain area. A thief can make three attempts to guess the four-digit pin code. The debit card is blocked after three unsuccessful attempts. What is the probability that the pin code of two or more debit cards is guessed correctly, assuming that each four-digit pin code is equally likely? (answer: 0.01017)

Problem 3.6. In the final of the World Series Baseball, two unevenly matched teams play a series consisting of at most seven games until one of the two teams has won four games. The probability that the weaker team will win any given game is 0.45 and the outcomes of the games are independent. What is the probability of the weaker team winning the final after six games? (answer: $0.27565 \times 0.45 = 0.1240$)

Problem 3.7. What is the expected number of values showing up two or more times when six fair dice are rolled? (answer: 1.579) *Hint*: use the linearity of expectation. Can you argue that the probability of some value showing up three or more times is about 0.37?

Problem 3.8. In an ESP-experiment a medium has to guess the correct symbol on each of 250 Zener cards. Each card has one of the five possible Zener symbols on it and each of the symbols is equally likely to appear. The medium will get $100 000 dollars if he gives 82 or more correct answers. What is the probability that the medium must be paid out? (answer: 1.36×10^{-6})

Problem 3.9. You enter a gambling house (stock market) with a bankroll of $100 and you are going to play a game with 10 sequential bets (investments). Each time you bet your whole current bankroll. A fair coin is tossed. Your current bankroll increases with 70% if heads appears and decreases with 50% if tails appears (an expected return of 10% for each bet!). Let G be your bankroll after the 10 bets. What is $P(G > 100)$? (answer: 0.3770). Can you explain why this probability is so small?

Problem 3.10. In the game "Lucky 10" twenty numbers are drawn from the numbers 1 to 80. You tick 10 numbers on the game form. What is the probability of matching 5 or more of the 20 numbers drawn? (answer: 0.0647)

Problem 3.11. A bowl contains 10 red and 15 white balls. You randomly pick without replacement one ball at a time until you have 5 red balls. What the probability that more than 10 picks are needed? (answer: 0.6626)

Problem 3.12. For a final exam, your professor gives you a list of 15 items to study. He indicates that he will choose eight for the actual exam. You will be required to answer correctly at least five of those. You decide to study 10 of the 15 items. What is the probability that you will pass the exam? (answer: $\frac{9}{11}$)

3.2 The Poisson distribution

A random variable X with $0, 1, \ldots$ as possible values is said to have a *Poisson distribution* with parameter $\lambda > 0$ if

$$P(X = k) = e^{-\lambda} \frac{\lambda^k}{k!} \quad \text{for } k = 0, 1, \ldots,$$

where $e = 2.71828\ldots$ is the Euler number. The expected value and the standard deviation of X are

$$E(X) = \lambda \quad \text{and} \quad \sigma(X) = \sqrt{\lambda}.$$

These formulas are proved in the appendix of this chapter.

The Poisson distribution is closely related to the binomial distribution. It is a good approximation to the probability distribution of the total number of successes in a *very large* number of independent trials each having a *very small* probability of success. This is easily seen for the probability of zero successes. In n physically independent trials each having a probability p of success, the probability of no success at all is

$$(1 - p) \times (1 - p) \times \cdots \times (1 - p) = (1 - p)^n.$$

By the approximation $e^{-x} \approx 1 - x$ for x close to 0, $(1 - p)^n$ can be approximated by $(e^{-p})^n = e^{-np}$ if p is very small. Therefore the probability of no success in n independent trials each having a very small success probability p is approximately equal to $e^{-\lambda}$ with $\lambda = np$. More generally, it can be proved that the binomial probability $\binom{n}{k} p^k (1-p)^{n-k}$ of getting exactly k successes in n independent trials tends to $e^{-\lambda} \lambda^k / k!$ as n gets very large and p very small such that np tends to a constant $\lambda > 0$, see the appendix of this chapter.

A very important fact is that only the product value $\lambda = np$ is relevant for the Poisson approximation to the binomial distribution with parameters n and p. You do not need to know the particular values of the number of trials and the success probability. It is enough to know what the expected (or average) value of the total number of successes is. This is an extremely useful property when it comes to practical applications. The physical background of the Poisson distribution, as a distribution of the total number of successes in a large number of trials each having a small probability of success, explains why this distribution has so many practical applications: the annual number of damage claims at insurance companies, the annual number of severe traffic accidents in a given region, the annual number of stolen credit cards, the annual number of fatal shark bites worldwide, etc.[17] Also, the Poisson distribution often provides a good description of many situations involving points randomly distributed in a bounded region in the plane or space.

The mathematical derivation of the Poisson distribution assumes that the trials are physically independent of each other, but, in many practical situations, the Poisson distribution also appears to give good approximations when there is a 'weak' dependence between the outcomes of the trials. The Poisson heuristic is especially useful for quickly arriving at good approximations in such problems for which it would otherwise be difficult to find exact solutions.

Example 3.4. There are 500 people present at a gathering. For the fun of it, the organizers have decided that all of those whose birthday is that day will receive a present. How many presents are needed to ensure a less than 1% probability of having too few presents?

Solution. Let the random variable X represent the number of individuals with a birthday on the day of the gathering. Leap year day, February 29, is discounted, and apart from that, it is assumed that every day of the year is equally likely as birthday. The distribution of X can then be modeled by a binomial distribution with parameters $n = 500$ and $p = \frac{1}{365}$. Calculating $P(X > k)$ reveals that

$$P(X > 4) = 0.0130 \quad \text{and} \quad P(X > 5) = 0.0028,$$

---

[17]Under rather weak conditions the Poisson distribution also applies under non-identical success probabilities of the trials.

Table 1: Binomial and Poisson probabilities

k	0	1	2	3	4	5	6
bin	0.2537	0.3484	0.2388	0.1089	0.0372	0.0101	0.0023
Poi	0.2541	0.3481	0.2385	0.1089	0.0373	0.0102	0.0023

and so five presents suffice for the meeting.

In this example, n is large and p is small such that the binomial distribution of X can be approximated by a Poisson distribution with expected value $\lambda = np = \frac{500}{365}$. Table 1 gives both the binomial probability and the Poisson probability that exactly k persons have a birthday on the day of the gathering for $k = 0, 1, \ldots, 6$. The Poisson probabilities are very close to the binomial probabilities.

The z-score test

A practically useful feature of the Poisson distribution is that the probability of a value more than three standard deviations removed from the expected value is very small (10^{-3} or smaller) when the expected value λ is sufficiently large. A rule of thumb is $\lambda \geq 25$. This feature is very useful for judging the value of all sorts of statistical facts reported in the media. In order to judge how exceptional a certain random outcome is, you measure how many standard deviations the outcome is removed from the expected value. This is called the *z-score test* in statistics. For example, suppose that in a given year the number of break-ins occurring in a given area increases more than 15% from an average of 64 break-ins per year to 75 break-ins. Since the z-score is $(75 - 11)/\sqrt{64} = 1.38$, the increase can be explained as a chance fluctuation and so there is no reason to demand the resignation of the police officer.

Example 3.5. The Pegasus Insurance Company has introduced a policy that covers certain forms of personal injury with a standard payment of \$100 000. On average, 100 claims per year lead to payment. There are many tens of thousands of policyholders. What can be said about the probability that more than 15 million dollars will have to be paid out in the space of a year?

Solution. In fact, every policyholder conducts a personal experiment in probability after purchasing this policy, which can be considered to be "successful" if the policyholder files a rightful claim during the ensuing year. In view of the many policyholders, there is a large number of independent probability experiments each having a very small probability of success. Therefore the Poisson model can be used. This model is entirely determined by the expected value of 100 for the number of rightful claims. Denoting by the random variable X the total number of claims that will be approved for payment during the year of coverage, the random variable X can be modeled by a Poisson distribution with parameter $\lambda = 100$. The probability of having to pay out more than 15 million dollars is given by $P(X > 150)$. Since $E(X) = 100$ and $\sigma(X) = 10$, a value of 150 claims lies five standard deviations above the expected value. Thus, without doing any further calculations, you can draw the conclusion that the probability of paying out more than 15 million dollars in the space of a year must be extremely small. The precise value of the probability $P(X > 150)$ is $1 - \sum_{k=0}^{150} e^{-100} \frac{100^k}{k!} = 1.23 \times 10^{-6}$. Not a probability the insurance executives need worry about.

For a binomially distributed random variable X with parameters n and p it is also true that almost all probability mass from the distribution lies within three standard deviations from the expected value when $np(1 - p)$ is sufficiently large. A rule of thumb for this is $np(1 - p) \geq 20$. A beer brewery once made brilliant use of this. In a television advertisement spot broadcast during the American Super Bowl final, 100 beer drinkers were asked to do a blind taste test comparing beer brewed by the sponsored brewery, and beer brewed by a competitor. The brilliance of the stunt is that the 100 beer drinkers invited to participate were regular drinkers of the brand made by the competitor. In those days, all brands of American beer tasted more or less the same, and most drinkers weren't able to distinguish between brands. The marketers of the sponsored beer could therefore be pretty sure that more than 35% of the participants in the stunt would prefer the sponsored beer over their regular beer (the target value of 35 is $(50 - 35)/5 = 3$ standard deviations below the expected value of 50 and so the binomial probability of falling below 35 is very small). This did, in fact, occur, and made quite an impression on the television audience. As another illustration, in Problem 3.8 the z-value for 82 or more good answers is $(82-50)/\sqrt{40} \approx 5.06$ and so you can conclude without any further calculations that the probability of payout is extremely small.

Problem 3.13. What is the Poisson approximation for the sought probability in Problem 3.5? (answer: 0.01019)

Problem 3.14. What is the probability of the jackpot falling in lotto 6/42 when 5 million tickets are randomly filled in? (answer: 0.6145)

Problem 3.15. In a coastal area, the average number of serious hurricanes is 3.1 per year. Use an appropriate probability model to calculate the probability of a total of more than 5 serious hurricanes in the next year. (answer: 0.0943)

Problem 3.16. The low earth orbit contains many pieces of space debris. It is estimated that an orbiting space station will be hit by space debris beyond a critical size and speed on average once in 400 years. Estimate the probability that a newly launched space station will not be penetrated in the first 20 years. (answer: 0.951)

Problem 3.17. Suppose r dice are simultaneously rolled each time. A roll in which each of the r dice shows up a six is a called a king's roll (generalization of de Méré's dice problem). For larger values of r, what is the probability of not getting a king's roll in $4 \times 6^{r-1}$ rolls of the r dice? (answer: $e^{-2/3} = 0.5134$)

Problem 3.18. In a particular rural area, postal carriers are attacked by dogs 324 times per year on average. Last year there were 379 attacks. Is this exceptional? (answer: yes, the z-score is 3.1)

3.3 The normal probability density

Many probabilistic situations are better described by a continuous random variable rather than a discrete random variable. Think of the annual rainfall in a certain area or the decay time of a radioactive particle. Calculations in probability and statistics are often greatly simplified by approximating the probability mass function of a discrete random variable by a continuous curve. As illustrated in Figure 3, the probability mass function of the binomial distribution with parameters n and p can very well be approximated by the graph of the continuous function

$$f(x) = \frac{1}{\sigma\sqrt{2\pi}} \, e^{-\frac{1}{2}(x-\mu)^2/\sigma^2}$$

with $\mu = np$ and $\sigma = \sqrt{np(1-p)}$ if n is sufficiently large, say $np(1-p) \geq 20$. This function $f(x)$ is called the *normal density function* (or the *Gaussian density function*). Likewise the probability mass function of the Poisson distribution with parameter λ can be approximated by the graph of the normal density function $f(x)$ with $\mu = \lambda$ and $\sigma = \sqrt{\lambda}$ if λ is sufficiently large, say $\lambda \geq 25$. In Figure 3, the parameters of the binomial and the Poisson distributions are $(n = 125, \, p = \frac{1}{5})$ and $\lambda = 25$.

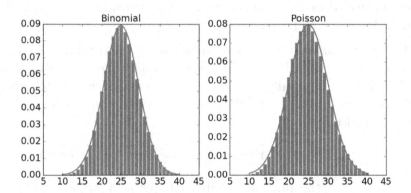

Figure 3: Normal approximation

You have now arrived at the normal distribution, which is the most important continuous distribution. A continuous random variable X is said to have a *normal distribution* with parameters μ and $\sigma > 0$ if

$$P(X \leq x) = \frac{1}{\sigma\sqrt{2\pi}} \int_{-\infty}^{x} e^{-\frac{1}{2}(y-\mu)^2/\sigma^2} \, dy \qquad \text{for } -\infty < x < \infty.$$

The notation $N(\mu, \sigma^2)$ is often used for a normally distributed random variable X with parameters μ and σ.

The normal density function $f(x) = \frac{1}{\sigma\sqrt{2\pi}} e^{-\frac{1}{2}(x-\mu)^2/\sigma^2}$ is the derivative of the probability distribution function $F(x) = P(X \leq x)$. The parameters μ and σ of the normal density function are the expected value and the standard deviation of X, that is,

$$\mu = \int_{-\infty}^{\infty} x f(x)\,dx \quad \text{and} \quad \sigma^2 = \int_{-\infty}^{\infty} (x - \mu)^2 f(x)\,dx.$$

The derivation of these formulas is beyond the scope of this book. The following remarks are made. The normal density function $f(x)$ is maximal for $x = \mu$ and is symmetric around the point $x = \mu$. The point $x = \mu$ is also the median of the normal probability distribution.[18] About 68.3% of the probability mass of a normally distributed random variable with expected value μ and standard deviation σ is between the points $\mu - \sigma$ and $\mu + \sigma$, about 95.4% between $\mu - 2\sigma$ and $\mu + 2\sigma$, and about 99.7% between $\mu - 3\sigma$ and $\mu + 3\sigma$. These facts are displayed in Figure 4 and will be explained below after having introduced the standard normal distribution.

The normal distribution has the important property that $aX + bY$ is normally distributed for any constants a and b if the random variables X and Y are normally distributed and independent of each other. The expected value μ and the standard deviation σ of $aX + bY$ are then equal to

$$\mu = aE(X) + bE(Y) \quad \text{and} \quad \sigma = \sqrt{a^2\sigma^2(X) + b^2\sigma^2(Y)}.$$

In particular, the random variable $aX + b$ is $N(a\mu + b, a^2\sigma^2)$ distributed if the random variable X is $N(\mu, \sigma^2)$ distributed.

The *standard normal distribution* is the normal distribution with expected value 0 and standard deviation 1. This distribution is usually denoted as the $N(0, 1)$ distribution. For a standard normally

---

[18]The median of a continuous random variable is defined as a point such that the random variable has 50% of its probability mass left from that point and 50% of its probability mass right from that point. It is noted that for a continuous random variable each individual point has probability mass zero.

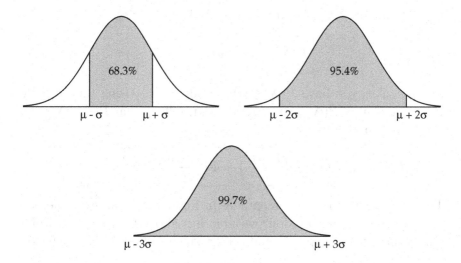

Figure 4: Characteristics of the normal density function

distributed random variable Z, the notation

$$\Phi(z) = P(Z \le z)$$

is used for the probability distribution function of Z. The function $\Phi(z)$ is given by the famous integral

$$\Phi(z) = \frac{1}{\sqrt{2\pi}} \int_{-\infty}^{z} e^{-\frac{1}{2}x^2} \, dx \quad \text{for any } z.$$

In statistics, $\Phi(z)$ is computed as $\Phi(z) = \frac{1}{2} + \frac{1}{2}\text{erf}(z/\sqrt{2})$, where $\text{erf}(x) = \frac{2}{\sqrt{\pi}} \int_{0}^{x} e^{-y^2} \, dy$ is the Gauss error function.

If X has an $N(\mu, \sigma^2)$ distribution, then, by $E(aX+b) = aE(X)+b$ and $\sigma^2(aX + b) = a^2\sigma^2(X)$, the *normalized* random variable

$$Z = \frac{X - \mu}{\sigma}$$

has a standard normal distribution. This is a very useful result for the calculation of the probabilities $P(X \le x)$. Writing

$$P(X \le x) = P\left(\frac{X - \mu}{\sigma} \le \frac{x - \mu}{\sigma}\right),$$

you see that $P(X \le x)$ can be calculated as

$$P(X \le x) = \Phi\left(\frac{x - \mu}{\sigma}\right).$$

In particular, you have $P(X \le \mu) = \Phi(0) = 0.5$. Using the formula $P(a < X \le b) = P(X \le b) - P(X \le a)$, it follows that $P(a < X \le b)$ can be calculated as

$$P(a < X \le b) = \Phi\left(\frac{b - \mu}{\sigma}\right) - \Phi\left(\frac{a - \mu}{\sigma}\right) \text{ for any } a < b.$$

This result explains the percentages in Figure 4. As an example, an $N(\mu, \sigma^2)$ distributed random variable has about 95.4% of its probability mass between $\mu - 2\sigma$ and $\mu + 2\sigma$, since $\Phi(2) - \Phi(-2) = 0.9545$.

As an illustration of the normal distribution, the length of Northern European boys who are born after a gestational period between 38 and 42 weeks has a normal distribution with an expected value of 50.9 cm and a standard deviation of 2.4 cm at birth. How exceptional is it that a boy at birth has a length of 48 cm? A quick answer to this question can be given by using the z-score test: a length of 48 cm is $\frac{50.9 - 48}{2.4} = 1.2083$ standard deviations below the expected value, and this is not exceptional. The undershoot probability $\Phi(-1.2083) = 0.1135$ corresponds to a z-score of -1.2083.

As said before, the normal distribution is the most important continuous distribution. Many stochastic situations in practice can be modeled with the help of the normal distribution. For example, the annual rainfall in a certain area, the cholesterol level of an adult male of a specific racial group, errors in physical measurements, the length of men in a certain age group, etc. Figure 5 displays a histogram of height measurements of a large number of men in a certain age group. A histogram divides the range of values covered by the measurements into intervals of the same width, and shows the proportion of the measurements in each interval. You see that the histogram has the characteristic bell-shaped form of the graph of the normal density function. Making the width of the intervals smaller and smaller and the number of observations larger and larger, the graph of the histogram changes into the graph of a normal density function. It is

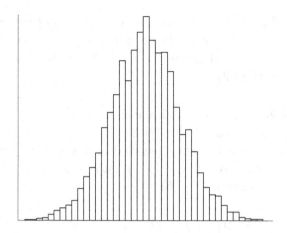

Figure 5: Histogram of height measurements

much simpler to work with the continuous normal density function than with the probability mass function underlying the histogram.

Problem 3.19. The annual rainfall in Amsterdam has a normal distribution with an expected value of 799.5 mm and a standard deviation of 121.4 mm. What is the probability of having more than 1 000 mm rainfall in Amsterdam next year? (answer: 0.0493)

Problem 3.20. Gestation periods of humans have a normal distribution with an expected value of 280 days and a standard deviation of 10 days. What is the percentage of births that are more than 15 days overdue? (answer: 0.0668)

Problem 3.21. The diameter of a 1 euro coin has a normal distribution with an expected value of 23.25 mm and a standard deviation of 0.10 mm. A vending machine accepts only 1 euro coins with a diameter between 22.90 mm and 23.60 mm. What is the probability that a 1 euro coin will not be accepted? (answer: 4.65×10^{-4})

Problem 3.22. The annual grain harvest in a certain area is normally distributed with an expected value of 15 000 tons and a standard deviation of 2 000 tons. In the past year the grain harvest was 21 500 tons. Is this exceptional? (answer: yes, the z-score is 3.075)

Problem 3.23. What is the standard deviation of the demand for a certain item if the demand has a normal distribution with an expected value of 100 and the probability of a demand exceeding 125 is 0.05? (answer: 15.2)

Problem 3.24. A stock return can be modeled by an $N(\mu, \sigma^2)$ distributed random variable. An investor believes that there is a 10% probability of a return below \$80 and a 10% probability of a return above \$120. What are the investor's estimates of μ and σ? (answer: $\mu = 100$ and $\sigma = 15.6$)

Problem 3.25. Verify that $P(|X - \mu| > k\sigma) = 2[1 - \Phi(k)]$ for any $k > 0$ if X is $N(\mu, \sigma^2)$ distributed.

3.4 Central limit theorem and the normal distribution

In this section the most famous theorem of probability and statistics is discussed. This theorem explains why many stochastic situations in practice can be modeled with the help of a normal distribution. As an empirical fact, a random variable that can be seen as the result of the sum of a large number of small independent random effects is approximately normally distributed. Mathematically, this result is expressed by the *central limit theorem*, which is the most celebrated theorem in probability and statistics. Loosely formulated,

> **if the random variables $X_1, X_2, \ldots, X_n$ are independent of each other and have each the same probability distribution with expected value μ and standard deviation σ, then the sum $X_1 + X_2 + \cdots + X_n$ has approximately a normal distribution with expected value $n\mu$ and standard deviation $\sigma\sqrt{n}$ if n is sufficiently large.**

Alternatively, it can be said: the sample mean $\frac{1}{n}(X_1 + X_2 + \cdots + X_n)$ has approximately a normal distribution with expected value μ and standard deviation $\frac{\sigma}{\sqrt{n}}$ if n is sufficiently large (use the fact that $E(aX) = aE(X)$ and $\sigma(aX) = a\sigma(X)$ for any constant $a > 0$).

Mathematically, for any value of x,

$$P(X_1 + X_2 + \cdots + X_n \leq x) \longrightarrow \Phi\left(\frac{x - n\mu}{\sigma\sqrt{n}}\right) \quad \text{as } n \to \infty,$$

$$P\left(\frac{X_1 + X_2 + \cdots + X_n}{n} \leq x\right) \longrightarrow \Phi\left(\frac{x - \mu}{\sigma/\sqrt{n}}\right) \quad \text{as } n \to \infty.$$

How large n should be depends on the shape of the probability distribution of the X_i: the more symmetric this distribution is, the sooner the normal approximation applies. The central limit theorem is extremely useful for both practical and theoretical purposes. To illustrate, consider Example 2.15 again. Using the fact that an $N(\alpha, \beta^2)$ distributed random variable has 95.4% of its probability mass between $\alpha - 2\beta$ and $\alpha + 2\beta$, there is a probability of about 95% that the net profit of the school after 500 bets will be between $250 - 2 \times 25 = 200$ dollars and $250 + 2 \times 25 = 300$ dollars.

The central limit theorem explains why the histogram of the probability mass function of a binomially distributed random variable with parameters n and p can be nicely approximated by the graph of a normal density with expected valued np and standard deviation $\sqrt{np(1-p)}$ if n is sufficiently large: a binomial random variable can be written as the sum $X_1 + \cdots + X_n$ of n independent random variables X_i with $P(X_i = 1) = p$ and $P(X_i = 0) = 1 - p$.

The central limit theorem has an interesting history. The first version of this theorem was postulated in 1738 by the French-born English mathematician Abraham de Moivre, who used the normal distribution to approximate the distribution of the number of heads resulting from many tosses of a fair coin. De Moivre's finding was far ahead of its time, and was nearly forgotten until the famous French mathematician Pierre Simon Laplace rescued it from obscurity in his monumental work *Théorie Analytique des Probabilités*, which was published in 1812. Laplace expanded De Moivre's finding by approximating the binomial distribution with the normal distribution. But as with De Moivre, Laplace's finding received little attention in his own time. It was not until the nineteenth century was at an end that the importance of the central limit theorem was discerned, when, in 1901, the Russian mathematician Aleksandr Lyapunov defined it in general terms and proved precisely how it worked mathematically.

The central limit theorem and the law of large numbers are two pillars of probability theory. The law of large numbers states that $\frac{1}{n}\sum_{k=1}^{n} X_k$ tends to $E(X)$ with probability one as n tends to infinity, and the central limit theorem enables you to give probabilistic error bounds on deviations of $\frac{1}{n}\sum_{k=1}^{n} X_k$ from $E(X)$ for finite but large n. These matters will come back in Chapter 5 on simulation, including the statistical concept of confidence interval.

Example 3.6. For an expedition with a duration of one and a half years, a number of spare copies of a particular filter must be taken along. The filter will be used daily. The lifetime of the filter has a continuous probability distribution with an expected value of one week and a standard deviation of half a week. Upon failure a filter is replaced immediately by a new one. How many filters should be taken along with the expedition in order to ensure that there will be no shortage with a probability of at least 99%?

Solution. Suppose that n filters are taken along for the expedition of 78 weeks. The probability of no shortage during the expedition is $P(X_1 + \cdots + X_n > 78)$, where X_i is the lifetime (in weeks) of the ith filter. The lifetimes of the filters are assumed to be independent of each other. Then, by the central limit theorem,

$$P(X_1 + \cdots + X_n > 78) = 1 - P(X_1 + \cdots + X_n \leq 78)$$
$$= 1 - P\left(\frac{X_1 + \cdots + X_n - n}{0.5\sqrt{n}} \leq \frac{78 - n}{0.5\sqrt{n}}\right) \approx 1 - \Phi\left(\frac{78 - n}{0.5\sqrt{n}}\right).$$

The requirement is that $1 - \Phi\big((78 - n)/(0.5\sqrt{n})\big) \geq 0.99$, and so you need the smallest value of n for which $\Phi\big((78 - n)/(0.5\sqrt{n})\big) \leq 0.01$. The solution of the equation $\Phi(x) = 0.01$ is $x = -2.326$ (the so-called 1% percentile[19]). Next you solve the equation

$$\frac{78 - z}{0.5\sqrt{z}} = -2.326.$$

The solution is $z = 88.97$. Thus 89 copies of the filter are needed.

---

[19]For any $0 < p < 1$, the $100p\%$ percentile z_p of the standard normal distribution is defined as the unique solution to $\Phi(x) = p$. For example, $z_{0.025} = -1.960$ and $z_{0.975} = 1.960$.

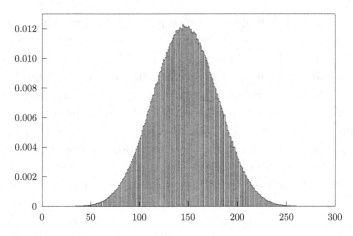

Figure 6: Probability histogram for $r = 6$ and $s = 49$

Statistical application

The central limit theorem also applies to the following situation in which you have weakly dependent random variables. Suppose that r distinct numbers are sequentially drawn from the numbers 1 up to s, one at a time and at random. Let the random variables X_i represent the ith number drawn. For reasons of symmetry each of the X_i has the same probability distribution, but they are not independent. If s is large and r is much smaller than s, the dependency between the X_i is weak. Then, it can be proved that $X_1 + \cdots + X_r$ is approximately $N(\mu, \sigma^2)$ distributed with

$$\mu = \frac{1}{2}r(s+1) \quad \text{and} \quad \sigma^2 = \frac{1}{12}r(s-r)(s+1).$$

This result can be used to demystify a widely advertised lottery system claimed to increase one's chances to win the lottery. As an example, let's take the lottery 6/49 in which six different numbers are randomly drawn from the numbers 1 to 49. The lottery system is based on the bell curve and fools lottery players by suggesting them to choose six different numbers that add up to a number between 117 and 183. If you take $r = 6$ and $s = 49$ in the above probability model, you get $\mu = 150$ and $\sigma = 32.8$. The normal curve has about

68% of its probability mass between $\mu - \sigma$ and $\mu + \sigma$. That's why the lottery system suggests players to choose six lottery numbers that add up to a number between $150 - 33 = 117$ and $150 + 33 = 183$. It is true that the sum of the six winning numbers will fall between 117 and 183 with a probability of about 68%, but this lottery system does not increase the player's odds of winning a prize. The advice completely neglects the fact that there are many more combinations of six numbers whose sum falls in the middle of the sum's distribution than there combinations of six numbers whose sum falls in the tail of the distribution. Figure 6 displays the simulated frequency of the sum of the six winning lottery numbers, based on one million draws. As you see, the probability histogram can accurately be approximated by a normal density function.

Problem 3.26. An insurance company has $20\,000$ policyholders. The amount claimed yearly by a policyholder has an expected value of \$150 and a standard deviation of \$750. Calculate an approximation for the probability that the total amount claimed in the coming year will exceed 3.3 million dollars. (answer: 0.0023)

Problem 3.27. Let the random variable H_n be the number of heads showing up in n tosses of a fair coin. What is the approximate distribution of $H_n - \frac{1}{2}n$ for large n? (answer: $N(0, (\frac{1}{2}\sqrt{n})^2)$)

Problem 3.28. In the random walk on the line, a drunkard takes each time a unit step to the right or to the left with equal probabilities, independently of the previous steps. Let D_n be the distance of the drunkard from the starting point after n steps. What is an approximation to $E(D_n)$ for n large? (answer: $\sqrt{2n/\pi}$).

Problem 3.29. The Nero Palace casino has a new, exciting gambling machine: the multiplying bandit. How does it work? The bandit has a lever or "arm" that the player may depress up to ten times. After each pull, an H (heads) or a T (tails) appears, each with probability $\frac{1}{2}$. The game is over as soon as heads appears or the player has pulled the arm ten times. The player wins $\$2^k$ if heads appears at the k pull for $1 \leq k \leq 10$, and wins \$1 500 otherwise. The stake for this game is \$15. What are the expected value and the standard deviation of the money payout for the player? (answer: $11.4648 and $64.1235) What is the approximate distribution of the casino's profit over $2\,500$ games? (answer: $N(8\,838, (3\,206)^2)$)

Problem 3.30. A new online casino has just opened and is making a promotional offer. Each of the first $2\,500$ online bets of $10 on red at roulette gets back $5 when the bet is lost and $20 when the bet is won. A bet on red is lost with probability $\frac{19}{37}$ and is won with probability $\frac{18}{37}$. Use the central limit theorem to approximate the probability that the casino will lose no more than $6\,500$ dollars on the promotional offer. (answer: 0.978)

3.5 The uniform and exponential probability densities

First, we discuss in more detail the tricky and subtle concept of probability density function. A random variable X is said to be *continuously distributed* with *probability density function $f(x)$* if

$$P(X \le x) = \int_{-\infty}^{x} f(y)\,dy \quad \text{for all real numbers } x,$$

where $f(x)$ is a non-negative function with $\int_{-\infty}^{\infty} f(x)\,dx = 1$ and $f(x)$ is continuous with the possible exception of a finite number of points.[20] The number $f(x)$ is not a probability but it is a relative measure for the likelihood that the random variable X will take on a value in the immediate neighborhood of (continuity) point x:

$$P(x < X \le x + \Delta x) \approx f(x)\Delta x \quad \text{for } \Delta x \text{ close to 0.}$$

This follows from $P(x < X \le x+\Delta x) = P(X \le x+\Delta x) - P(X \le x)$ and the fact that $f(x)$ is the derivative of $P(X \le x)$ (if $g(x)$ is the derivative of $G(x)$, then $G(x + \Delta x) - G(x) \approx g(x)\Delta x$ for Δx close to 0). The density function $f(x)$ specifies how the probability mass of the continuously distributed random variable X is smeared out, as were it liquid mass, over the range of the possible values of X. Note

---

[20]In general, you find the probability density function of a continuous random variable X by determining $F(x) = P(X \le x)$ and differentiating $F(x)$.

that a density function can have values larger than 1, though it must integrate to 1.

In view of the interpretation of $f(x)\Delta x$ for Δx very small, it is reasonable to define the expected value and variance of X as

$$E(X) = \int_{\infty}^{\infty} x f(x)\, dx \quad \text{and} \quad \text{var}(X) = \int_{-\infty}^{\infty} \left[x - E(X)\right]^2 f(x)\, dx,$$

assuming that the integrals are well-defined and are finite. Using the formula $P(a < X \leq b) = P(X \leq b) - P(X \leq a)$ and the integral representation of $P(X \leq x)$, you have

$$P(a < X \leq b) = \int_{a}^{b} f(x)\, dx \quad \text{for } a < b.$$

The integral $\int_{a}^{b} f(x)\, dx$ is the area under the graph of the density function $f(x)$ between the points a and b. This area goes to 0 if a tends to b. Thus each individual point has probability mass *zero* for the random variable X, and so $P(a < X < b) = P(a < X \leq b) = P(a \leq X < b) = P(a \leq X \leq b)$.

3.5.1 The uniform density function

Let the random variable X be a random point in the finite interval (a, b), where a random point means that the probability of X falling in a subinterval of (a, b) is proportional to the length of the subinterval (think of throwing blindly a dart with an infinitely thin point on the interval (a, b)). The proportionality constant must be $\frac{1}{b-a}$ (why?). In particular, for any $a < x < b$, the probability of X falling in the interval (a, x) is $(x - a)/(b - a)$. Thus

$$P(X \leq x) = \frac{x - a}{b - a} \quad \text{for } a < x < b$$

with $P(X \leq x) = 0$ for $x \leq a$ and $P(X \leq x) = 1$ for $x \geq b$. The derivative of $P(X \leq x)$ on (a, b) is

$$f(x) = \frac{1}{b - a} \quad \text{for } a < x < b.$$

Defining $f(x) = 0$ for $x \notin (a, b)$, it follows that $f(x)$ is the probability density function of the random point X. This density function is called the *uniform density function* on (a, b). The probability mass of X is evenly spread out over the interval (a, b). The uniform density function underlies the so-called random numbers in computer simulation. It is a matter of simple algebra to verify that

$$E(X) = \frac{a+b}{2} \quad \text{and} \quad \text{var}(X) = \frac{1}{12}(b-a)^2.$$

Beta density

The uniform density on $(0, 1)$ is a special case of the beta density. The class of beta densities is much used in Bayesian analysis. To introduce the beta density, imagine that you have a biased coin with an unknown probability of coming up heads. Suppose that you model your ignorance about the true value of this probability by a random variable having the uniform density on $(0, 1)$ as probability density. In other words, the prior density $f(p)$ of your belief about the probability of coming up heads is $f(p) = 1$ for $0 < p < 1$. How does your belief about the probability of coming up heads change when you have tossed the coin n times with heads coming up s times and tails $r = n - s$ times? Denote by $f(p \mid s \text{ heads})$ the posterior density of your belief about the probability of coming up heads. This density satisfies the Bayes formula

$$f(p \mid s \text{ heads}) = \frac{f(p)\, L(s \text{ heads} \mid p)}{\int_0^1 f(\theta)\, L(s \text{ heads} \mid \theta)\, d\theta},$$

where the likelihood $L(s \text{ heads} \mid p)$ is the probability of getting s heads and r tails in $r + s$ tosses of the coin when the probability of coming up heads is p. You are asked to take this formula for granted, see also Section 2.4. The essence of the formula is that the posterior density is proportional to the product of the prior density and the likelihood function. In the coin-tossing example the likelihood function is given by a binomial probability:

$$L(s \text{ heads} \mid p) = \binom{r+s}{s} p^s\, (1-p)^r \quad \text{for } 0 < p < 1.$$

This leads to

$$f(p \mid s \text{ heads}) = \frac{f(p) \binom{r+s}{s} p^s\, (1-p)^r}{\int_0^1 f(\theta) \binom{r+s}{s} \theta^s\, (1-\theta)^r\, d\theta},$$

and so, with $f(p) = 1$ for $0 < p < 1$, you get the posterior density

$$f(p \mid s \text{ heads}) = \frac{p^s (1-p)^r}{\int_0^1 \theta^s (1-\theta)^r \, d\theta} \qquad \text{for } 0 < p < 1.$$

This density is a beta density. The beta (α, β) density with parameters $\alpha > 0$ and $\beta > 0$ has the interval $(0,1)$ as its range and is defined by

$$\boxed{\frac{1}{B(\alpha, \beta)} x^{\alpha-1} (1-x)^{\beta-1} \qquad \text{for } 0 < x < 1,}$$

where the normalizing constant $B(\alpha, \beta) = \int_0^1 x^{\alpha-1} (1-x)^{\beta-1} \, dx$ with $B(\alpha, \beta) = (\alpha-1)!(\beta-1)!/(\alpha+\beta-1)!$ for the case that α and β are integer-valued. Using advanced calculus, it can be shown that the expected value and the variance of a random variable X with a beta (α, β) density are given by

$$\boxed{E(X) = \frac{\alpha}{\alpha + \beta} \quad \text{and} \quad \text{var}(X) = \frac{\alpha\beta}{(\alpha+\beta)^2(\alpha+\beta+1)}.}$$

The uniform density on $(0, 1)$ is the beta (α, β) density with $\alpha = \beta = 1$. A closer look at the above derivation shows that you would have found the beta $(\alpha + s, \beta + r)$ density as posterior density for the probability of coming up heads if you had taken the beta (α, β) density as prior density for this probability. As an illustration, assuming the uniform density as prior density for the probability of coming up heads and tossing the coin 10 times, then the posterior density becomes the beta $(8, 4)$ density when heads appears $s = 7$ times and tails $r = 3$ times. If another 10 tosses are done and these tosses result in 6 heads and 4 tails, the posterior density becomes the beta $(14, 8)$ density.

3.5.2 The exponential density function

The density function of continuous random variable X is said to be an *exponential density function* with parameter $\lambda > 0$ if

$$\boxed{f(x) = \lambda e^{-\lambda x} \qquad \text{for } x \geq 0}$$

and $f(x) = 0$ for $x < 0$. A random variable X with this density can only take on positive values. The exponential probability distribution is often used to model the time until the occurrence of a rare event (e.g., serious earthquake, the decay of a radioactive particle). Figure 7 gives the histogram of a large number of observations of the time until decay of a radioactive particle. An exponential density function can indeed be very well fitted to this histogram.

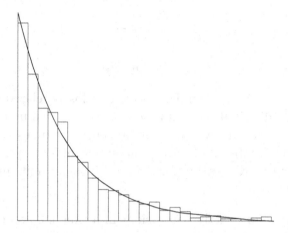

Figure 7: A histogram for decay times

The probability distribution function $P(X \le x)$ is (verify!)

$$P(X \le x) = 1 - e^{-\lambda x} \quad \text{for } x \ge 0.$$

A basic formula in integral calculus is $\int_0^\infty x^k e^{-\lambda x}\, dx = \frac{k!}{\lambda^{k+1}}$ for $k = 0, 1, \dots$ and $\lambda > 0$. Using this formula, you can readily verify that

$$E(X) = \int_0^\infty x f(x)\, dx = \frac{1}{\lambda}, \ \ \text{var}(X) = \int_0^\infty \left(x - \frac{1}{\lambda}\right)^2 f(x)\, dx = \frac{1}{\lambda^2}.$$

A characteristic property of the exponential random variable X is its *lack of memory*. That is, for any $s > 0$,

$$P(X > s + x \mid X > s) = P(X > x) \quad \text{for all } x > 0.$$

In other words, imagining that X represents the lifetime of an item, the residual life of the item has the same exponential distribution as the original lifetime, regardless of how long the item has been already in use ('used is as good as new'). The proof is very simple. Since $P(X > v) = e^{-\lambda v}$ for all $v > 0$, you find for any $s > 0$ that

$$P(X > s + x \mid X > s) = \frac{P(X > s + x \text{ and } X > s)}{P(X > s)}$$

$$= \frac{P(X > s + x)}{P(X > s)} = \frac{e^{-\lambda(s+x)}}{e^{-\lambda s}} = e^{-\lambda x} \quad \text{for all } x > 0.$$

Poisson process

The continuous exponential distribution is closely related to the discrete Poisson distribution. Suppose that radioactive material emits particles one at a time, where the interoccurrence times are independent random variables having a same exponential density function. Letting λ be the average number of emissions per unit time, then it can be shown that

$P(\text{there will occur exactly } k \text{ emissions in a time interval of length } t)$

$$= e^{-\lambda t} \frac{(\lambda t)^k}{k!} \quad \text{for } k = 0, 1, \ldots,$$

independently of the number of emissions that occurred before that time interval. That's why this counting process of the number of particles emitted is called a *Poisson process*. This process is one of the most important random processes in probability theory.

As shown by the Russian mathematician Alexsandr Khinchin (1894–1959), the Poisson process arises only if the following conditions are satisfied:

• Events (e.g., decay events) occur one at a time, that is, two or more events in a very small time interval is practically impossible.
• The numbers of events in non-overlapping time intervals are independent of one another (no after-effects).
• The probability distribution of the number of events occurring during any finite time interval depends only on the length of the interval and not on its position on the time axis.

The physical background of the Poisson process explains why this process can be used to model random processes in many areas such as inventory control, maintenance and reliability, queueing, and telecommunications.

Example 3.7. In a traditional match between two university soccer teams the lengths of time between goals are independent of each other and have an exponential distribution with an expected value of 30 minutes. The playing time of the match is 90 minutes. What is the probability of having three or more goals during the match? What is the probability that exactly two goals will be scored in the first half of the match and exactly one goal in the second half?

Solution. Goals are scored at a rate of $\lambda = \frac{1}{30}$ per minute. The number of goals scored during the match has a Poisson distribution with an expected value of $\lambda \times 90 = 3$. Therefore the probability of having three or more goals during the match is given by

$$1 - \sum_{k=0}^{2} e^{-3} \frac{3^k}{k!} = 0.5768.$$

By the memoryless property of the exponential distribution, the number of goals scored in the first half of the match and the number of goals scored in the second half of the match are independent of each other and have each a Poisson distribution with an expected value of $\lambda \times 45 = 1.5$. Thus the probability that exactly two goals will be scored in the first half of the match and exactly one goal in the second half is given by

$$e^{-1.5} \frac{1.5^2}{2!} \times e^{-1.5} \frac{1.5}{1!} = 0.0840.$$

Problem 3.31. A satellite has a lifetime that is exponentially distributed with an expected value of 15 years. The satellite is in use for already 12 years? What are the expected value and the standard deviation of the residual lifetime of the satellite (answer: 15 years) What is the probability that the satellite will survive for another 10 years? (answer: 0.5134)

Problem 3.32. The lifetime X of an item is exponentially distributed with mean $\frac{1}{\lambda}$. Verify that, for any x,

$$\boxed{P(x < X \le x + \Delta x \mid X > x) \approx \lambda \Delta x \text{ for } \Delta x \text{ close to zero.}}$$

What is the probabilistic meaning of this result? (answer: constant failure rate)

Problem 3.33. You go by bus to work each day. It takes 5 minutes to walk from home to the bus stop. In order to get to work on time, you must take a bus no later than 7:45 a.m. The independent interarrival times of the busses are exponentially distributed with a mean of 10 minutes. What is the latest time you must leave home in order to be on time for work with a probability of at least 0.95? (answer: 7:10 a.m.)

Problem 3.34. In a hospital five babies are born per 24 hours on average. It is reasonable to model the times between the arrivals of the babies by independent exponential random variables each having an expected value of $\frac{24}{5}$ hours. What is the probability that more than two babies will be born between twelve o'clock midnight and six o'clock in the morning? (answer: 0.1315)

Problem 3.35. In a video game with a time slot of fixed length T, signals occur according to a Poisson process with rate λ, where $T > \frac{1}{\lambda}$. In the time slot you can push a button only once. You win if at least one signal occurs in the time slot and you push the button at the occurrence of the last signal. Your strategy is to let pass a fixed time s with $0 < s < T$ and push the button upon the first occurrence of a signal (if any) after time s. What value of s maximizes your probability of winning the game and what is your maximum win probability? (answer: $s = T - \frac{1}{\lambda}$ and $\frac{1}{e}$)

3.6 The bivariate normal density[21]

Multivariate normal distributions are the most important joint distributions of two or more random variables. This section briefly discusses the case of two random variables.

A random vector (X, Y) is said to have a *standard bivariate normal distribution* with parameter ρ if

$$P(X \leq x \text{ and } Y \leq y) = \int_{-\infty}^{x} \int_{\infty}^{y} \varphi(v, w)\, dw \, dv \text{ for } -\infty < x, y < \infty,$$

where the *standard normal density function* $\varphi(x, y)$ is given by

$$\varphi(x, y) = \frac{1}{2\pi\sqrt{1 - \rho^2}}\, e^{-\frac{1}{2}\left(x^2 - 2\rho xy + y^2\right)/\left(1 - \rho^2\right)} \qquad \text{for all } x \text{ and } y.$$

The parameter ρ is a constant with $-1 < \rho < 1$. The density function $\varphi(x, y)$ allows for the interpretation: the probability that the pair (X, Y) will take on a value in a small rectangle around the point (x, y) with side lengths Δx and Δy is approximately $\varphi(x, y)\Delta x \Delta y$.

A bivariate normal density function has elliptical contours, that is, for each $c > 0$, the set $\{(x, y) : \varphi(x, y) = c\}$ is an ellipse. A bivariate normal density function is displayed in Figure 8.

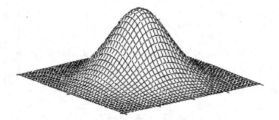

Figure 8: A bivariate normal density

For a pair (X, Y) having a standard bivariate normal distribution with parameter ρ, it is a matter of integral calculus to show that

---

[21]This section be skipped without loss of continuity.

- both X and Y are $N(0,1)$ distributed[22]
- the correlation coefficient of X and Y is ρ.

More generally, the vector (X,Y) is said to have a bivariate normal density with parameters $(\mu_X, \mu_Y, \sigma_X^2, \sigma_Y^2, \rho)$ if the standardized vector $\left(\frac{X-\mu_1}{\sigma_1}, \frac{Y-\mu_2}{\sigma_2}\right)$ has the standard bivariate normal density with parameter ρ. Then, by taking the partial derivatives of $P(X \leq x$ and $Y \leq y) = \int_0^{(x-\mu_1)/\sigma_1} \int_0^{(y-\mu_2)/\sigma_2} \varphi(v,w)\,dw\,dv$ with respect to x and y, the joint density $f_{X,Y}(x,y)$ of X and Y is

$$\frac{1}{2\pi\sigma_1\sigma_2\sqrt{1-\rho^2}}e^{-\frac{1}{2}\left[\left(\frac{x-\mu_1}{\sigma_1}\right)^2 - 2\rho\left(\frac{x-\mu_1}{\sigma_1}\right)\left(\frac{y-\mu_2}{\sigma_2}\right) + \left(\frac{y-\mu_2}{\sigma_2}\right)^2\right]/(1-\rho^2)}.$$

Also, it can be shown that X and Y have $N(\mu_X, \sigma_X^2)$ and $N(\mu_Y, \sigma_Y^2)$ densities and ρ is the correlation coefficient of X and Y. A characteristic property of the bivariate normal density is that independence of X and Y and uncorrelatedness $(\rho = 0)$ are equivalent.

In general, a linear combination of normally distributed random variables is not normally distributed. However, if (X,Y) has a bivariate normal distribution with parameters $(\mu_X, \mu_Y, \sigma_X^2, \sigma_Y^2, \rho)$, then $aX + bY + c$ has a normal distribution for any constants a, b and c. The expected value μ and the variance σ^2 of this distribution are

$$\boxed{\mu = a\mu_X + b\mu_Y + c \ \text{ and } \ \sigma^2 = a^2\sigma_X^2 + b^2\sigma_Y^2 + 2ab\rho\sigma_X\sigma_Y,}$$

noting that $\mathrm{var}(aX + bY + c) = a^2\mathrm{var}(X) + b^2\mathrm{var}(Y) + 2ab\mathrm{cov}(X,Y)$.

In applications, an important concept is the *conditional density function* of Y given that $X = x$. This density function is denoted by $f_Y(y \mid x)$ and is uniquely determined by the relation

$$\boxed{f_{X,Y}(x,y) = f_X(x)f_Y(y \mid x),}$$

---

[22]The proof is based on $P(X \leq x) = \int_{-\infty}^x \left(\int_{-\infty}^\infty \varphi(v,w)\,dw\right)dv$. Differentiation gives that the density of X is $\int_{-\infty}^\infty \varphi(x,w)\,dw$. Next, by the decomposition $\varphi(x,w) = (1/\sqrt{2\pi})\,e^{-\frac{1}{2}x^2} \times \left(1/(\sqrt{1-\rho^2}\sqrt{2\pi})\right)e^{-\frac{1}{2}(w-\rho x)^2/(1-\rho^2)}$, you get that $(1/\sqrt{2\pi})e^{-\frac{1}{2}x^2}$ is the marginal density of X. Similarly, the density of Y. The decomposition of $\varphi(x,y)$ is also the key to the verification of $E(XY) = \rho$. This gives $\mathrm{cov}(X,Y) = \rho$, and so the correlation coefficient $\rho(X,Y) = \rho$.

where $f_{X,Y}(x, y)$ is the joint density of the vector (X, Y) with parameters $(\mu_X, \mu_Y, \sigma_X^2, \sigma_Y^2, \rho)$ and $f_Y(y)$ is the density of the $N(\mu_Y, \sigma_Y^2)$ distributed random variable Y. This formula is the continuous analogue of $P(V = v$ and $W = w) = P(V = v)P(W = w \mid V = v)$ for discrete random variables V and W. The conditional density function $f_Y(y \mid x)$ can be shown to be a normal density function whose expected value μ_{cY} and variance σ_{cY}^2 are given by

$$\mu_{cY} = \mu_Y + \rho\frac{\sigma_Y}{\sigma_X}(x - \mu_X) \quad \text{and} \quad \sigma_{cY}^2 = \sigma_Y^2(1 - \rho^2).$$

Problem 3.36. A statistics class has two exams and the scores of the students on the exams 1 and 2 follow a bivariate normal distribution with parameters $\mu_1 = 75$, $\mu_2 = 65$, $\sigma_1 = 12$, $\sigma_2 = 15$ and $\rho = 0.7$. Take a randomly chosen student. What is the probability that the score on exam 1 is 80 or more? (answer: 0.3385) What is the probability that total score over the two exams will exceed 150? (answer: 0.3441) What is the probability that the student will do better on exam 2 than on exam 1? (answer: 0.1776) What is the probability that the score on exam 2 will be over 80 given that the score on exam 1 is 80? (answers: 0.1606)

Random walk application

The random walk model is a useful probability model in the physical sciences. Let us formulate this model in terms of a particle moving on the two-dimensional plane. The particle starts at the origin $(0, 0)$. In each step the particle travels a unit distance in a randomly chosen direction between 0 and 2π, independently of the other steps. What is the joint density function of the (x, y)-coordinates of the position of the particle after n steps?

Let the random variable Θ denote the direction taken by the particle in any step. In each step the x-coordinate of the position of the particle changes with an amount that is distributed as $\cos(\Theta)$ and the y-coordinate with an amount that is distributed as $\sin(\Theta)$. The continuous random variable Θ has a uniform distribution on $(0, 2\pi)$. Let X_k and Y_k be the changes of the x-coordinate and the y-coordinate of the position of the particle in the kth step. Then the position of the particle after n steps can be represented by the random vector (S_{n1}, S_{n2}), where

$$S_{n1} = X_1 + \cdots + X_n \quad \text{and} \quad S_{n2} = Y_1 + \cdots + Y_n.$$

You will not be surprised to learn that a two-dimensional central limit theorem can be applied to the vector (S_{n1}, S_{n2}). Letting

$$\mu_1 = E(X_1), \ \mu_2 = E(Y_1), \ \sigma_1^2 = \sigma^2(X_1), \ \sigma_2^2 = \sigma^2(Y_1), \ \rho = \rho(X_1, Y_1),$$

the two-dimensional central limit theorem says that the density function of the normalized vector $\left(\frac{S_{n1}-n\mu_1}{\sigma_1\sqrt{n}}, \frac{S_{n2}-n\mu_2}{\sigma_2\sqrt{n}}\right)$ is approximately given by the standard bivariate normal density with parameter ρ. In the particular case of the random walk, you have

$$\mu_1 = \mu_2 = 0, \ \sigma_1 = \sigma_2 = \frac{1}{\sqrt{2}} \quad \text{and} \quad \rho = 0.$$

The derivation of this result is simple and instructive. The random variable Θ has the uniform density function $f(\theta) = \frac{1}{2\pi}$ for $0 < \theta < 2\pi$. Applying the substitution rule gives

$$\mu_1 = E[\cos(\Theta)] = \int_0^{2\pi} \cos(\theta) f(\theta) \, d\theta = \frac{1}{2\pi} \int_0^{2\pi} \cos(\theta) \, d\theta = 0.$$

In the same way, $\mu_2 = 0$. Using the substitution rule again,

$$\sigma_1^2 = E\left[\cos^2(\Theta)\right] = \int_0^{2\pi} \cos^2(\theta) f(\theta) \, d\theta = \frac{1}{2\pi} \int_0^{2\pi} \cos^2(\theta) \, d\theta.$$

In the same way, $\sigma_2^2 = \frac{1}{2\pi} \int_0^{2\pi} \sin^2(\theta) \, d\theta$. Invoking the celebrated formula $\cos^2(\theta) + \sin^2(\theta) = 1$ from trigonometry, we obtain $\sigma_1^2 + \sigma_2^2 = 1$. Thus, for reasons of symmetry, $\sigma_1^2 = \sigma_2^2 = \frac{1}{2}$. Finally, $\rho = 0$ since $\text{cov}(X_1, Y_1) = E[\cos(\Theta)\sin(\Theta)] = \frac{1}{2\pi} \int_0^{2\pi} \cos(\theta)\sin(\theta) \, d\theta = 0$.

It can now be concluded that (S_{n1}, S_{n2}) has approximately the bivariate normal density function

$$f_n(x, y) = \frac{1}{\pi n} e^{-(x^2+y^2)/n} \quad \text{for large } n.$$

To conclude, define the random variable $D_n = \sqrt{S_{n1}^2 + S_{n2}^2}$ as the distance between the origin and the position of the particle after n steps. Then, by a two-dimensional substitution rule,

$$E(D_n) \approx \frac{1}{\pi n} \int_{-\infty}^{\infty} \int_{-\infty}^{\infty} \sqrt{x^2 + y^2}\, e^{-(x^2+y^2)/n}\, dx\, dy \quad \text{for large } n.$$

This integral can be evaluated by the standard technique of converting Cartesian coordinates (x, y) to polar coordinates (r, θ) by $x = r\cos(\theta)$ and $y = r\sin(\theta)$ with $dx\, dy = r\, dr\, d\theta$. Omitting the details, it follows after tedious algebra that

$$E(D_n) \approx \frac{1}{2}\sqrt{\pi n} \quad \text{for large } n.$$

The above results can be extended to the random walk in dimension three. Then, $E(D_n) \approx \sqrt{\frac{8n}{3\pi}}$ for n large.

3.7 The chi-square test

The chi-square test (χ^2 test) is one of the most useful statistical tests. It is used to test whether data were generated from a particular probability distribution. The test can also be used to judge whether data have been manipulated to make the observed frequencies closer to the expected ones.

Suppose you want to find out whether the probability mass function $p_1, \ldots, p_r$ fits a random sample of observations obtained for a repeatable chance experiment with a finite number of possible outcomes $O_1, \ldots, O_r$. To introduce the chi-square test, denote by the random variable N_j the number of times that outcome O_j will appear in n physically independent repetitions of the chance experiment, where outcome O_j has probability p_j. The random variable N_j has a binomial distribution with parameters n and p_j, and so the expected value of N_j is np_j. By the principle of least squares, it is reasonable to consider a test statistic of the form $\sum_{j=1}^{r} w_j(N_j - np_j)^2$ for appropriately chosen weights w_j. It turns out that the choice $w_j = \frac{1}{np_j}$ yields a statistic with a tractable distribution. Thus the so-called *chi-square statistic* is defined by

$$D = \sum_{j=1}^{r} \frac{(N_j - np_j)^2}{np_j}.$$

The probability distribution of the statistic D is difficult to compute. However, the discrete probability distribution of D can be very accurately approximated by a tractable continuous distribution when np_j is sufficiently large for all j, say $np_j \geq 5$ for all j (in order to achieve this, it might be necessary to pool some data groups). Then,

$$P(D \leq x) \approx P(\chi^2_{r-1} \leq x) \quad \text{for } x \geq 0,$$

where the continuous random variable χ^2_{r-1} is distributed as the sum of the squares of $r - 1$ independent $N(0, 1)$ distributed random variables. The probability distribution of χ^2_{r-1} is called the chi-square distribution with $r - 1$ degrees of freedom. Its expected value is $E(\chi^2_{r-1}) = r - 1$. It should be pointed out that the above approximation for the chi-square statistic D assumes that the probabilities p_j are not estimated from the data but are known beforehand; if you have to estimate one or more parameters to get the probabilities p_j, you must lower the number of degrees of freedom of the chi-square distribution by one for every parameter estimated from the data.

How do you apply the chi-square test in practice? Using the data that you have obtained for the chance experiment in question, you calculate the numerical value d of the test statistic D for these data. The (subjective) judgment whether the probability mass function $p_1, \ldots, p_r$ fits the data depends on the value of the probability $P(D \leq d)$. All this will be illustrated in the following two examples.

Example 3.8. Somebody claims to have rolled a fair die 1 200 times and to have found that the outcomes 1, 2, 3, 4, 5, and 6 occurred 196, 202, 199, 198, 202, and 203 times. Do you believe these results?

Solution. The reported frequencies are very close to the expected frequencies. Since the expected value and the standard deviation of the number of rolls with outcome j are $1\,200 \times \frac{1}{6} = 200$ and $\sqrt{1\,200 \times (1/6) \times (5/6)} = 12.91$ for all j, you should be suspicious about the reported results. You can substantiate this with the chi-square test. The chi-square statistic D takes on the value

$$\frac{1}{200}\left[(196-200)^2+(202-200)^2+(199-200)^2+(198-200)^2\right.$$
$$\left.+(202-200)^2+(203-200)^2\right]=0.19.$$

The value 0.19 lies far below the expected value 5 of the chi-square distribution with $6-1=5$ degrees of freedom. The probability $P(D\leq 0.19)$ is approximated by

$$P(\chi_5^2\leq 0.19)=0.00078.$$

The simulated value of $P(D\leq 0.19)$ is 0.00083 (four million simulation runs). The very small probability for the test statistic indicates that the data are indeed most likely fabricated.

Example 3.9. A total of 64 matches were played during the World Cup soccer 2010 in South Africa. There were 7 matches with zero goals, 17 matches with 1 goal, 13 matches with two goals, 14 matches with three goals, 7 matches with four goals, 5 matches with five goals, and 1 match with seven goals. Does a Poisson distribution fit closely these data?

Solution. In this example you must first estimate the parameter λ of the hypothesized Poisson distribution. The estimate for λ equals $\frac{1}{64}\left(17\times 1+13\times 2+14\times +7\times 4+5\times 5+0\times 6+1\times 7\right)=\frac{145}{64}$. In order to satisfy the requirement that each data group should have an expected size of at least 5, the matches with 5 or more goals are aggregated, and so six data groups are considered. If a Poisson distribution with expected value $\lambda=\frac{145}{64}$ applies, then the expected number of matches with exactly j goals is $64\times e^{-\lambda}\lambda^j/j!$ for $j=0,1,\ldots,4$ and the expected number of matches with 5 or more goals is $64\times(1-\sum_{j=0}^{4}e^{-\lambda}\lambda^j/j!)$. These expected numbers have the respective values 6.641, 15.046, 17.044, 12.872, 7.291, and 5.106. Thus the value of the chi-square test statistic D is given by

$$\frac{(7-6.641)^2}{6.641}+\frac{(17-15.046)^2}{15.046}+\frac{(13-17.044)^2}{17.044}+\frac{(14-12.872)^2}{12.872}$$
$$+\frac{(7-7.291)^2}{7.291}+\frac{(6-5.106)^2}{5.106}=1.500.$$

Since the parameter λ has been estimated from the data, the test statistic D has approximately a chi-square distribution with $6 - 1 - 1 = 4$ degrees of freedom. The probability $P(\chi_4^2 \geq 1.500) = 0.827$. Thus the Poisson distribution gives an excellent fit to the data.

Problem 3.37. In a classical study on the distribution of 196 soldiers kicked to death by horses among 14 Prussian cavalry corps over the 20 years 1875 to 1894, the data are as follows. In 144 corps-years no deaths occurred, 91 corps-years had one death, 32 corps-years had two deaths, 11 corps-years had three deaths, and 2 corps-years had four deaths. Does a Poisson distribution fit the data? (answer: yes, $P(\chi_2^2 \geq 1.952) = 0.377)$[23]

Problem 3.38. In a famous physics experiment done by Rutherford, Chadwick and Ellis in 1920, the number of α-particles emitted by a piece of radioactive material were counted during $2\,608$ time intervals of each 7.5 seconds. Denoting by O_j the number of intervals with exactly j particles, the observed data are $O_0 = 57$, $O_1 = 203$, $O_2 = 383$, $O_3 = 525$, $O_4 = 532$, $O_5 = 408$, $O_6 = 273$, $O_7 = 139$, $O_8 = 45$, $O_9 = 27$, $O_{10} = 10$, $O_{11} = 4$, $O_{12} = 0$, $O_{13} = 1$, and $O_{14} = 1$. Do the observed frequencies conform to Poisson frequencies? (answer: yes, $P(\chi_{10}^2 \geq 12.961) = 0.226)$

---

[23]This study was done by the Russian statistician Ladislaus von Bortkiewicz (1868–1931), who first discerned and explained the importance of the Poisson distribution in his book *Das Gesetz der Kleinen Zahlen.* The French mathematician Siméon-Denis Poisson (1781–1840) himself did not recognize the huge practical importance of the distribution that would later be named after him.

Appendix: Poisson and binomial probabilities

This appendix shows that the Poisson distribution can be obtained as the limiting distribution of the binomial probabilities

$$b_k = \binom{n}{k} p^k (1-p)^{n-k} \quad \text{for } k = 0, 1, \ldots, n.$$

If n tends to infinity and p to zero such that np goes to a constant $\lambda > 0$, then the binomial probabilities b_k converge to the Poisson probabilities

$$p_k = e^{-\lambda} \frac{\lambda^k}{k!} \quad \text{for } k = 0, 1, \ldots .$$

To avoid technicalities, the limiting result is only proved for the case that np is kept fixed on the value λ ($p = \frac{\lambda}{n}$). The proof then goes as follows: write $\binom{n}{k} p^k (1-p)^{n-k}$ as

$$\binom{n}{k} \left(\frac{\lambda}{n}\right)^k \left(1 - \frac{\lambda}{n}\right)^{n-k} = \frac{n!}{k!\,(n-k)!} \frac{\lambda^k}{n^k} \frac{(1-\lambda/n)^n}{(1-\lambda/n)^k}$$

$$= \frac{\lambda^k}{k!} \left(1 - \frac{\lambda}{n}\right)^n \left[\frac{n!}{n^k\,(n-k)!}\right] \left(1 - \frac{\lambda}{n}\right)^{-k}.$$

The next step in the proof is to fix k and to analyze separately the terms in the last expression. The first term $\frac{\lambda^k}{k!}$ is left untouched. The second term $\left(1 - \frac{\lambda}{n}\right)^n$ tends to $e^{-\lambda}$ if n tends to infinity. The third term can be written as:

$$\frac{n!}{n^k\,(n-k)!} = \frac{n\,(n-1)\cdots(n-k+1)}{n^k}$$

$$= \left(1 - \frac{1}{n}\right) \left(1 - \frac{2}{n}\right) \cdots \left(1 - \frac{k-1}{n}\right).$$

Thus, for fixed k, the third term tends to 1 if n tends to infinity. For fixed k, the last term $\left(1 - \frac{\lambda}{n}\right)^{-k}$ also tends to 1 if n tends to infinity. This completes the proof.

Let X be a Poisson distributed random variable with parameter λ. Then,

$$E(X) = \lambda \quad \text{and} \quad \sigma(X) = \sqrt{\lambda}.$$

To prove this, use the relations

$$\sum_{k=0}^{\infty} k e^{-\lambda} \frac{\lambda^k}{k!} = \lambda \quad \text{and} \quad \sum_{k=0}^{\infty} k(k-1) e^{-\lambda} \frac{\lambda^k}{k!} = \lambda^2,$$

which are readily verified by using $k! = k \times (k-1)! = k(k-1) \times (k-2)!$ and $\sum_{j=0}^{\infty} e^{-\lambda} \lambda^j / j! = 1$. This gives $E(X) = \lambda$ and $E[X(X-1)] = \lambda^2$, and so $E(X^2) = \lambda^2 + \lambda$ and $\sigma^2(X) = \lambda$.

Thinned Poisson distribution

Let X be a Poisson distributed random variable with parameter λ. Think of X as the number of events that will occur in a given period. Imagine that each event is marked with probability p_1 and is left unmarked with probability $p_2 = 1 - p_1$, independently of any other event. Let the random variable X_1 be the number of marked events and $X_2 = X - X_1$ be the number of unmarked events. Then, X_i is Poisson distributed with parameter λp_i for $i = 1, 2$, and, more surprisingly, X_1 and X_2 are independent of each other.

The proof is simple and is another demonstration of the close ties between Poisson and binomial probabilities. First, note that

$$P(X_1 = j \text{ and } X_2 = k) = P(X_1 = j \text{ and } X = j + k)$$
$$= P(X = j + k) P(X_1 = j \mid X = j + k),$$

using the formula $P(A \text{ and } B) = P(B) P(A \mid B)$. This gives

$$P(X_1 = j \text{ and } X_2 = k) = e^{-\lambda} \frac{\lambda^{j+k}}{(j+k)!} \binom{j+k}{j} p_1^j p_2^k \quad \text{for } j, k = 0, 1 \ldots.$$

Writing $\binom{j+k}{j} = (j+k)! / (j! \, k!)$ and rearranging the terms, you get

$$P(X_1 = j \text{ and } X_2 = k) = e^{-\lambda p_1} \frac{(\lambda p_1)^j}{j!} \times e^{-\lambda p_2} \frac{(\lambda p_2)^k}{k!} \quad \text{for } j, k = 0, 1 \ldots.$$

The proof is now always done. Summing $P(X_1 = j \text{ and } X_2 = k)$ over k gives the Poisson distribution $P(X_1 = j) = e^{-\lambda p_1}(\lambda p_1)^j / j!$ for $j = 0, 1, \ldots$. Similarly, $P(X_2 = k) = e^{-\lambda p_2}(\lambda p_2)^k / k!$ for $k = 0, 1, \ldots$. Since $P(X_1 = j \text{ and } X_2 = k) = P(X_1 = j) P(X_2 = k)$ for all j, k, the random variables X_1 and X_2 are independent.

Chapter 4
Real-Life Examples of Poisson Probabilities

The Poisson distribution was introduced in Section 3.2. It is appropriate for modeling situations in which there is a very large number of opportunities for an event to occur, but each with a very low chance of occurrence. The Poisson distribution is as nearly important as the normal distribution. Whereas the normal distribution has two parameters, namely the expected value and the standard deviation, the Poisson distribution depends only on its expected value (the standard deviation is the square root of the expected value). In this chapter a number of surprising applications of the Poisson distribution in everyday life are given.

4.1 Fraud in a Canadian lottery

In the Canadian province of Ontario, a strong suspicion arose at a certain point that winning lottery tickets were repeatedly stolen by lottery employees from people who had their lottery ticket checked at a point of sale. These people, mostly the elderly, were then told that their ticket had no prize and that it could go into the trash. The winning ticket was subsequently surrendered by the lottery ticket seller, who pocketed the cash prize. The ball started to roll when an older participant – who always entered the same numbers on his ticket – found out that in 2001 a prize of $250 000 was taken from him by a sales point employee.

How do you prove a widespread fraud in the lottery system? Statistical analysis by Jeffrey Rosenthal, a well-known Canadian professor of probability, showed the fraud. The Poisson distribution played a major role in the analysis. How was the analysis done? Rosenthal worked together with a major Canadian TV channel. Investigation by the TV channel, making an appeal to the freedom of information act, revealed that in the period 1999–2006 there were 5 713 big prizes ($50 000 or more) of which 200 prizes were won by lottery ticket sellers. Can this be explained as a fluke of chance? To answer this question, you need to know how many people are working at points of sale of the lottery. There were 10 300 sales outlets in Ontario and research by the TV channel led to an estimated average of 3.5 employees per point of sale, or about 36 thousand employees in total. The lottery organization fought this number and came up with 60 thousand lottery ticket sellers. You also need to know how much the average lottery ticket seller spends on the purchase of lottery tickets compared to the average adult inhabitant of Ontario. The estimate by the TV channel was that the average expenditure on lottery tickets taken over all lottery ticket sellers was about 1.5 times as large as the average expenditure on lottery tickets taken over all 8.9 million adult residents of Ontario.

Let's now calculate the probability that the lottery sellers will win 200 or more of the 5 713 big prizes when there are 60 thousand lottery ticket sellers with an expenditure factor of 1.5. In that case the expected number of winners of big prizes among the lottery ticket sellers can be estimated as

$$5\,713 \times \frac{60\,000 \times 1.5}{8\,900\,000} = 57.$$

In view of the physical background of the Poisson distribution – the probability distribution of the total number of successes in a very large number of independent trials each having a very small probability of success – it is plausible to use the Poisson distribution to model the number of winners among the lottery ticket sellers. The Poisson distribution has the nice feature that its standard deviation is the square root of its expected value. Moreover, nearly all the probability mass of the Poisson distribution lies within three standard deviations from the expected value when the expected value is not too small, see Section 3.2. Two hundred winners among the lottery ticket sellers lies

$$\frac{200 - 57}{\sqrt{57}} \approx 19$$

standard deviations above the expected value. The probability corresponding to a z-score of 19 is inconceivably small (on the order of 10^{-49}) and makes clear that there is large-scale fraud in the lottery.

The lottery organization objected to the calculations and came with new figures. Does the conclusion of large-scale fraud change for the rosy-tinted figures of 101 000 lottery ticket sellers with an expenditure factor of 1.9? Then you get the estimate

$$5\,713 \times \frac{101\,000 \times 1.9}{8\,900\,000} \approx 123.$$

for the expected number of winners under the lottery ticket sellers. Two hundred winners is still

$$\frac{200 - 123}{\sqrt{123}} \approx 7$$

standard deviations above the expected value. A z-score of 7 has also a negligibly small probability (on the order of 10^{-7}) and cannot be explained as a chance fluctuation. It could not be otherwise that there was large-scale lottery fraud at the sales points of lottery tickets. This suspicion was also supported by other research findings. The investigations led to a great commotion. Headings rolled and the control procedures were adjusted to better protect the customer. The stores' ticket checking machines must now be viewable by customers, and make loud noises to indicate wins. Customers are now required to sign their names on their lottery tickets before redeeming them, to prevent switches.

4.2 Bombs over London in World War II

A famous application of the Poisson model is the statistical analysis of the distribution of hits of flying bombs (V-1 and V-2 missiles) in London during the second World War. The British authorities were anxious to know if these weapons could be accurately aimed at a particular target, or whether they were landing at random. If the missiles were in fact only randomly targeted, the British could simply disperse important installations to decrease the likelihood of their being hit. An area of 36 square kilometers in South London was divided into 576 small regions of 250 meter wide by 250 meter long, and the number of hits in each region was determined. There were 229 regions with zero hits, 211 regions with one hit, 93 regions with two hits, 35 regions with three hits, 7 regions with four hits, 1 region with five hits, and 0 regions with six or more hits. The 576 regions were struck by $229 \times 0 + 211 \times 1 + 93 \times 2 + 35 \times 3 + 7 \times 4 + 1 \times 5 = 535$ bombs and so the average number of hits per region was

$$\lambda = \frac{535}{576} = 0.9288.$$

If it can be made plausible that the number of hits per region closely follows a Poisson distribution with an expected value of 0.9288, then it can be safely concluded that the missiles landed at random, in view of properties of the two-dimensional Poisson process on the plane. You would expect

$$V_k = 576 \times e^{-0.9288} \frac{0.9228^k}{k!}$$

regions with exactly k hits for $k = 0, 1, \ldots$ if the number of hits per sector is approximately Poisson distributed with an expected value of 0.9288. Calculating V_k for $k = 0, 1, \ldots, 5$, you get

$V_0 = 227.5, V_1 = 211.3, V_2 = 98.1, V_3 = 30.4, V_4 = 7.1$ and $V_5 = 1.3$.

You see that the observed relative frequencies 229, 211, 93, 35, 7 and 1 for the numbers of hits are each very close to these theoretical relative frequencies corresponding to a Poisson distribution (this can be formalized with the chi-square test from Section 3.7). It could be concluded that the distribution of hits in the South London area was much like the distribution of hits when a flying bomb was to fall on any of the equally sized regions with the same probability, independently of the other flying bombs. The statistical analysis convinced the British military that the bombs struck at random and had no advanced aiming ability.

4.3 Winning the lottery twice

The following item was reported in the February 14, 1986 edition of *The New York Times*: "A New Jersey woman wins the New Jersey State Lottery twice within a span of four months." She won the jackpot for the first time on October 23, 1985 in the 6/39 lottery. Then she won the jackpot in the new 6/42 lottery on February 13, 1986. In the r/s lottery r different numbers are randomly drawn from the numbers 1 to s and you win the jackpot if you have correctly predicted all six winning numbers. Lottery officials of New Jersey State Lottery declared that the probability of winning the jackpot twice in one lifetime is approximately one in 17.1 trillion. What do you think of this statement? The claim made in this statement is easily challenged. The officials' calculation proves correct only in the extremely farfetched case scenario of a given person submitting one ticket for the 6/39 lottery and one ticket for the 6/42 lottery just one time in his/her life. In this case, the probability of getting all six numbers right, both times, is equal to

$$\frac{1}{\binom{39}{6}} \times \frac{1}{\binom{42}{6}} = \frac{1}{1.71 \times 10^{13}}.$$

But the event of someone winning the jackpot twice is far from miraculous when you consider a very large number of people who play the lottery for many weeks. The explanation is the *law of truly large numbers*: any event with a nonzero probability will eventually occur when it is given enough opportunity to occur. Let's illustrate this lottery principle by considering the many 6/42 lotteries in the world that have two draws in each week. Suppose that each of 50 million people fills in five tickets for each draw. Then the probability of one of them winning the jackpot at least twice in the coming four years is close to 1. The calculation of this probability is based on the Poisson distribution, and goes as follows. The probability of winning the jackpot in a particular week when filling in five tickets is

$$\frac{5}{\binom{42}{6}} = 9.531 \times 10^{-7}.$$

In view of the physical background of the Poisson distribution, it is reasonable to model the number of times that a given player will win a jackpot in the next $4 \times 52 \times 2 = 416$ draws of a 6/42 lottery by a Poisson distribution with expected value

$$\lambda_0 = 416 \times \frac{5}{\binom{42}{6}} = 3.965 \times 10^{-4}.$$

For the next 416 draws, this means that

P(a particular player will win the jackpot two or more times)
$= 1 - e^{-\lambda_0} - e^{-\lambda_0}\lambda_0 = 7.859 \times 10^{-8}.$

Next, using again the physical background of the Poisson distribution, we can conclude that the number of people under the 50 million mark, who win the jackpot two or more times in the coming four years is Poisson distributed with expected value

$$\lambda = 50\,000\,000 \times (7.859 \times 10^{-8}) = 3.93.$$

Thus the probability that at some point in the coming four years at least one of the 50 million players will win the jackpot two or more times can be given as

$$1 - e^{-\lambda} = 0.980.$$

A probability very close to 1! A few simplifying assumptions are used to make this calculation, such as the players choose their six-number sequences randomly. This does not influence the conclusion that it may be expected once in a while, within a relatively short period of time, that *someone* will win the jackpot two times.

4.4 Santa Claus and a baby whisperer

In 1996, the James Randi Educational Foundation was founded by James Randi, a former top magician who fought and exposed mockery and pseudo-sciences. The goal of the foundation was to make the public and the media aware of the dangers associated with the performances of psychic mediums. James Randi offered a $1 million prize to anyone who could demonstrate psychic abilities. Obviously, this had to be demonstrated under verifiable test conditions, which would be agreed on beforehand. For example, someone like Uri Geller who claimed to be able to bend spoons without applying force could not bring his own spoons. Different mediums took up the challenge but nobody succeeded. The 'baby whisperer' Derek Ogilvie was one of the mediums who accepted the challenge. This medium claimed to be capable of extrasensory distant observations. He was allowed to choose a child with whom he thought he would have telepathic contact, and he was subjected to the following test. The medium was shown ten different toys that would be given to the child one after the other, in random order, out of sight of the medium. The child was taken to an isolated chamber, and each time the child received a toy, the medium was asked to say what toy it was. If the medium was right six or more times, he would win one million dollars. What is the probability of six or more correct answers?

The problem is in fact a variation of the Santa Claus problem: at a Christmas party, each one of a group of children brings a present, after which the children draw lots randomly to determine who gets which present. What is the probability that none of the children will wind up with their own present?[24] This probability can be easily obtained by the Poisson heuristic. Assume that there are n children at the party and imagine that the children are numbered as $1, 2, \ldots, n$. The Santa Claus problem can be formulated within the framework of a sequence of n trials. In the ith trial, a lot is drawn by the child having number i. Let's say that a trial is successful if the child draws the lot for his/her own present. Then, the success probability of each trial has the same value

$$\frac{(n-1)!}{n!} = \frac{1}{n}$$

(and so the order in which lots are drawn does not matter). Thus the expected value of the number of successes is $n \times \frac{1}{n} = 1$, regardless of the value of n. The outcomes of the trials are not independent of each other, but the dependence is 'weak' if n is sufficiently large. The success probability $\frac{1}{n}$ is small for n large. Then, as noted before in Section 3.2, you can use the Poisson heuristic for the probability distribution of the total number of successes. This probability

---

[24]The Santa Claus problem and its variations boil down to the following combinatorial problem. Take a random permutation of the integers $1, 2, \ldots, n$. What is the probability that none of the integers keeps its original position?

distribution is then approximated by a Poisson distribution with an expected value of 1. Thus, since a 'success' means that the child gets his/her own present, you get for $k = 1, 2, \ldots, n$,

$$P(\text{exactly } k \text{ children will get their own present}) \approx \frac{e^{-1}}{k!}.$$

Numerical investigations reveal that this is a remarkably good approximation for $n \geq 10$ (the first seven decimals of the approximate values agree with the exact values already for n as large as 10). In particular, taking $k = 0$, the probability that none of the children will get their own present is about $\frac{1}{e} = 0.36787\ldots$, or, about 36.8%, regardless of the number of children.

Going back to the ESP experiment with the medium, James Randi was in very little danger of having to cough up the loot. The probability of six or more correct answers for random guessing is practically equal to the Poisson probability $1 - \sum_{k=0}^{5} \frac{e^{-1}}{k!} = 0.0006$, or, about 0.06%. The medium perhaps thought, beforehand, that five correct guesses was the most likely outcome, and a sixth correct guess on top of that wasn't that improbable, so, why not go for it. In the test he had only one correct answer.

4.5 Birthdays and 500 Oldsmobiles

In 1982 the organizers of the Quebec Super Lotto decided to use a fund of unclaimed winnings to purchase 500 Oldsmobiles, which would be raffled off as a bonus prize among the 2.4 million lottery subscribers in Canada. They did this by letting a computer pick 500 times a random number from the 2.4 million registration numbers of the subscribers. To the lottery officials' astonishment, they were contacted by one subscriber claiming to have won two Oldsmobiles. The lottery had neglected to program the computer not to choose the same registration number twice. The probability of a pre-specified subscriber winning the car two times is indeed astronomically small, but not so the case of the probability that, out of 2.4 million subscribers, there will be someone whose number appears at least twice in the list of 500 winning numbers. The latter event has a probability of about 5%. That is quite a small probability, but not a negligible probability. How can we calculate the probability of 5%? To do so, let's translate the lottery problem into a birthday problem on a planet with $d = 2\,400\,000$ days in the year and a randomly formed group of $m = 500$ aliens. What is the probability that two or more aliens share a birthday, assuming that each day is equally likely as birthday? This probability can be accurately approximated by

$$1 - e^{-\frac{1}{2}m(m-1)/d}.$$

This formula is easily obtained by the Poisson heuristic. There are $\binom{m}{2} = \frac{1}{2}m(m-1)$ different combinations of two aliens. Each combination has the same success probability of $\frac{1}{d}$ that the two aliens have a common birthday. In other words, you have a very large number of $\frac{1}{2}m(m-1)$ trials, each having the same tiny success probability $\frac{1}{d}$. The dependence between the trials is very weak and so the probability of no success is approximately equal to the Poisson probability $e^{-\lambda}$, where $\lambda = \frac{1}{2}m(m-1) \times \frac{1}{d}$ is the expected number of successful trials. This verifies the above approximation formula.

Identifying the Oldsmobiles with the aliens and the registration numbers with the birthdays, and substituting $d = 2\,400\,000$ and $m = 500$ in the approximation formula, you get the approximate value 0.0507 for the probability of winning two or more Oldsmobiles by some lottery subscriber. The approximate value is very accurate and agrees with the exact value in the first five decimals. The exact formula for the probability of two or more matching birthdays is

$$1 - \frac{d(d-1)\cdots(d-m+1)}{d^m},$$

as can be seen by the same arguments as used for the classical birthday problem with 365 equally likely birthdays, see Section 1.2. In this birthday problem, 23 people suffice to have a fifty-fifty match probability under the assumption of equally likely birthdays. In reality birthdays are not uniformly distributed throughout the year, but follow a seasonal pattern. However, for birth frequency variation as occurring in reality, the match probability is very insensitive to deviations from uniform birth rates. Empirical studies have been done that confirm this fact. For example, during the 2014 World Cup soccer championship, 32 national teams of 23 players each took part. It turned out that 18 of those teams had at least one double birthday (double birthdays for 15 teams at the 2018 World Cup soccer). The 2019 World Cup women's soccer had 24 teams of 23 players each and had 10 teams with at least one double birthday.

Problem 4.1. The thirteen cards of a particular suit are taken from a standard deck of 52 playing cards and are thoroughly shuffled. A dealer turns over the cards one at a time, calling out "ace, two, three, ..., king". A match occurs when the card turned over matches the rank called out by the dealer as he turns it over. What is the probability of a match occurring? (answer: $\frac{1}{e} = 0.3679$)

Problem 4.2. A blindfolded person is tasting ten different wines. Beforehand, he is informed of the names of the participating wineries, but is not told the order in which the ten wines will be served. The person may name each winery just once. During the taste-test, he succeeds in identifying five of the ten wineries correctly. Do you think this person is a wine connoisseur? (answer: yes, $P(\geq 5 \text{ successes by pure guessing}) = 3.66 \times 10^{-3}$)

Problem 4.3. You are sitting in a class of 15 students. The teacher randomly hands out a graded exam to each student. What is the probability that you will be the only student to get his/her own exam back? (answer: 0.0245)

Problem 4.4. What is the Poisson approximation for the probability that two or more aliens in the birthday problem from Section 4.5 have a birthday within one day from each other? (answer: $1 - e^{-\frac{3}{2}m(m-1)/d}$)

Problem 4.5. What is the Poisson approximation for the probability that no two cards of the same face value (two aces, for example) will succeed one another in a well-shuffled deck of 52 playing cards? (answer: e^{-3})

Problem 4.6. What is the Poisson approximation for the probability that the same combination of five numbers will appear two or more times in the next 100 draws of the lotto 5/32? (answer: 0.0243)

Chapter 5
Monte Carlo Simulation and Probability

Computer simulation is a natural partner for probability and imitates a concrete probability situation on the computer. In this chapter you will see how simulation works and how you can simulate many probability problems with relatively simple tools.[25] You will notice that simulation is not a simple gimmick, but requires mathematical modeling and algorithmic thinking. The emphasis is on the modeling behind computer simulation, not on the programming itself.

5.1 Introduction

Monte Carlo simulation is a powerful probabilistic analysis tool, widely used in both engineering fields and non-engineering fields. It is named after the famous gambling hot spot, Monte Carlo, in the Principality of Monaco. Monte Carlo simulation was initially used to solve neutron diffusion problems in atomic bomb research at Los Alamos National Laboratory in 1944. From the time of its introduction during World War II, Monte Carlo simulation has remained one of the most-utilized mathematical tools in scientific practice. And in addition to that, it has also functioned as a very useful tool for adding an extra dimension to the teaching and learning of probability. It may help students gain a better understanding of probabilistic

---

[25] The law of large numbers is the mathematical basis for the application of computer simulation to solve probability problems. The probability of a given event in a chance experiment can be estimated by the relative frequency of occurrence of the event in a very large number of repetitions of the experiment.

Table 2: Simulation results for 100 000 coin tosses

n	$H_n - \frac{1}{2}n$	f_n	n	$H_n - \frac{1}{2}n$	f_n
10	1	0.6000	5 000	-9.0	0.4982
25	1.5	0.5600	7 500	11	0.5015
50	2	0.5400	10 000	24	0.5024
100	2	0.5200	15 000	40	0.5027
250	1	0.5040	20 000	91	0.5045
500	-2	0.4960	25 000	64	0.5026
1 000	10	0.5100	30 000	78	0.5026
2 500	12	0.5048	100 000	129	0.5013

ideas and to overcome common misconceptions about the nature of 'randomness'. As an example, a key concept such as the law of large numbers can be made to come alive before one's eyes by watching the results of many simulation trials. The nature of this law is best illustrated through the coin-toss experiment. The law of large numbers says that the percentage of tosses to come out heads will be as close to 50% as you can imagine, provided that the number of coin tosses is large enough. But how large is large enough? Experiments have shown that the relative frequency of heads may continue to deviate significantly from 0.5 after many tosses, though it tends to get closer and closer to 0.5 as the number of tosses gets larger and larger. The convergence to the value 0.5 typically occurs in a rather erratic way. The course of a game of chance, although eventually converging in an average sense, is a whimsical process. To illustrate this, a simulation run of 100 000 coin tosses was made. Table 2 summarizes the results of this particular simulation study; any other simulation experiment will produce different numbers. The statistic $H_n - \frac{1}{2}n$ gives the observed number of heads minus the expected number after n tosses and the statistic f_n gives the observed relative frequency of heads after n tosses. It is worthwhile to take a close look at the results in the table. You see that the realization of the relative frequency, f_n, indeed approaches the true value of the probability in a rather irregular manner and converges more slowly than most of us would expect intuitively. That's why you should be suspicious of the

outcomes of simulation studies that consist of only a small number
of simulation runs, see also Section 5.4.

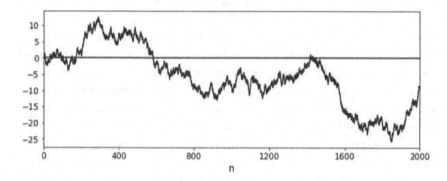

Figure 9: A random walk of 2 000 coin tosses

The law of large numbers does not imply that the absolute difference between the actual number of heads and the expected number
should oscillate close to zero. It is even typical for the coin-toss experiment that this difference has a tendency to become larger and
larger and to grow proportionally with the square root of the number
of tosses, whereby returns to 0 become rarer and rarer as the number
of coin tosses gets larger and larger. This is illustrated in Figure 9
which displays a simulated realization of the random walk describing
the actual number of heads minus the expected number over 2 000
coin tosses. The mathematical explanation of the growing oscilla-
tions displayed in the figure is provided by the central limit theorem
together with the square root law: the actual number of heads minus the expected number after n tosses is approximately normally
distributed with expected value 0 and standard deviation $\frac{1}{2}\sqrt{n}$ for n
large. The coin-toss experiment is full of surprises that clash with
intuitive thinking. Unexpectedly long sequences of either heads or
tails can occur ('local clusters' of heads or tails are absorbed in the
average). If you don't believe this, convince yourselves with simulation. Simulation can reveal interesting and surprising patterns.

Monte Carlo simulation is not only a very useful tool for helping
students to gain a better understanding of probabilistic ideas and

to overcome common misconceptions about the nature of "randomness." Simulation also enables you to get quick answers to specific probability problems or to check analytical solutions. For example, what is the probability that any two adjacent letters are different when the eleven letters of the word Mississippi are put in random order? Seemingly a simple probability problem, but it turns out that this combinatorial probability problem is difficult to solve analytically. In computational probability it is often beforehand not clear whether a probability problem easily allows for an analytical solution. Many probability problems are too difficult or too time-consuming to solve exactly, while a simulation program is easily written. Monte Carlo simulation can also be used to settle disagreement on the correct answer to a particular probability problem. It is easy to make mistakes in probability, so checking answers is important. Take the famous Monty Hall problem. This probability puzzle raised a lot of discussion about its solution. Paul Erdös, a world famous mathematician, remained unconvinced about the correct solution of the problem until he was shown a computer simulation confirming the correct result. In the Monty Hall problem, a contestant in a TV game show must choose between three doors. An expensive car is behind one of the three doors, and gag prizes are behind the other two. He chooses a door randomly, appealing to Lady Luck. Then the host opens one of the other two doors and, as promised beforehand, takes a door that conceals a gag prize. With two doors remaining unopened, the host now asks the contestant whether he wants to remain with his choice of door, or whether he wishes to switch to the other remaining door. What should the contestant do? Simulating this game is a convincing approach to show that it is better to switch, which gives a win probability of $\frac{2}{3}$.

5.2 Simulation tools

Simple tools often suffice for the simulation of probability problems. This section discusses first the concept of random generator. Next several useful simulation tools are presented. These tools include methods to generate a random point inside a bounded region and a random permutation of a finite set of objects. The simulations tools will be illustrated in the next section.

5.2.1 Random number generators

In the simulation of probability models, access to random numbers is of crucial importance. A *random number generator*, as it is called, is indispensable. A random number generator produces random numbers between 0 and 1 (excluding the values 0 and 1). For a 'truly' random number generator, it is as if fate falls on a number between 0 and 1 by pure coincidence. A random number between 0 and 1 is characterized by the property that the probability of the number falling in a subinterval of $(0, 1)$ is the same for each interval of the same length and is equal to the length of the interval. A truly random number can take on any possible value between 0 and 1. A random number from $(0, 1)$ enables you to simulate, for example, the outcome of a single toss of a fair coin without actually tossing the coin: if the generated random number is between 0 and 0.5 (the probability of this is 0.5), then the outcome of the toss is heads; otherwise, the outcome is tails. Producing random numbers is not as easily accomplished as it seems, especially when they must be generated quickly, efficiently, and in massive amounts. Even for simple simulation experiments the required amount of random numbers runs quickly into the hundreds of thousands or higher.[26] Generating a very large amount of random numbers on a one-time only basis, and storing them up in a computer memory, is practically infeasible. But there is a solution to this kind of practical hurdle that is as handsome as it is practical.

Instead of generating *truly* random numbers, a computer can generate so-called *pseudo random numbers*, and it achieves this through a nonrandom procedure. This idea comes from the famous Hungarian-American mathematician John von Neumann (1903–1957) who made very important contributions not only to mathematics but also to physics and computer science. The procedure for a pseudo random number generator is iterative by nature and is determined by a suitably chosen function f. Starting with a seed number z_0, numbers $z_1, z_2, \ldots$ are successively generated by $z_1 = f(z_0), z_2 = f(z_1)$, and

---

[26]In earlier times creative methods were sometimes used to generate random numbers. Around 1920 crime syndicates in New York City's Harlem used the last five digits of the daily published U.S. treasure balance of the American Treasury to generate the winning numbers for their illegal 'Treasury Lottery'.

so on. The function f is referred to as a pseudo random number generator and it must be chosen such that the sequence $\{z_i\}$ is statistically indistinguishable from a sequence of truly random numbers. The output of function f must be able to stand up to a great many statistical tests for 'randomness'.

The first pseudo random number generators were the so-called multiplicative congruential generators. Starting with a positive integer z_0, the z_i are generated by $z_i = az_{i-1}$ (modulo m) for $i = 1, 2, \ldots,$ where a and m are carefully chosen positive integers, e.g. $a = 16\,807$ and $m = 2^{31} - 1$. For this particular choice, the sequence $\{z_i\}$ repeats itself after $m - 1$ steps and thus the cycle length is m. The number z_i determines the random number u_i by $u_i = \frac{z_i}{m}$.

The newest pseudo random number generators do not use the multiplicative congruential scheme. In fact, they do not involve multiplications or divisions at all. These generators are very fast, have incredibly long periods before they repeat the same sequence of random numbers, and provide high-quality pseudo random numbers. In software tools you will find not only the so-called Christopher Columbus generator with a cycle length of about 2^{1492} (at ten million pseudo random numbers per second, it will take more than 10^{434} years before the sequence of numbers will repeat!), but you will also find the Mersenne twister generator with a cycle length of $2^{19937} - 1$. This generator would probably take longer to cycle than the entire future existence of humanity. It has passed numerous tests for randomness, including tests for uniformity of high-dimensional strings of numbers. The modern generators are needed in Monte Carlo simulations requiring huge masses of pseudo random numbers, as is the case in applications in physics and financial engineering.

In the sequel we omit the additive 'pseudo' and simply speak of random numbers and random number generator.

5.2.2 Simulating a random number from a finite interval

How do you choose randomly a number between two given numbers a and b with $a < b$? To do so, you first use the random number generator to get a random number u between 0 and 1. Next you find a random number x between a and b as

$$x = a + (b - a)u.$$

Verify yourselves that $0 < u < 1$ implies $a < a + (b - a)u < b$.

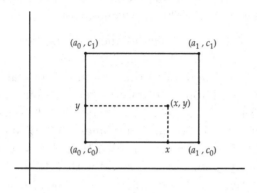

Figure 10: Simulating a random point inside a rectangle

The above procedure can be directly used to generate a random point inside a rectangle. Let (a_0, c_0), (a_1, c_0), (a_0, c_1) and (a_1, c_1) be the four corner points of the rectangle, see Figure 10. You first use the random number generator to get two random numbers u_1 and u_2 from $(0, 1)$. Then the random point (x, y) inside the rectangle is

$$x = a_0 + (a_1 - a_0)u_1 \text{ and } y = c_0 + (c_1 - c_0)u_2.$$

5.2.3 Simulating a random integer from a finite range

How do you choose randomly an integer from the integers $1, 2, \ldots, M$? To do so, you first use the random number generator to get a random number u between 0 and 1. Then a random integer k is

$$k = 1 + \text{int}(M \times u).$$

The function $\text{int}(x)$ rounds the number x to the nearest integer k that is smaller than or equal to x. That is,

$$\text{int}(x) = k \quad \text{if } k \le x < k + 1$$

for an integer k. More generally, a random integer k from the integers $a, a + 1, \ldots, b$ is obtained as

$$k = a + \text{int}\big((b - a + 1) \times u\big).$$

To illustrate, let's simulate the outcome of a single roll of a fair die ($M = 6$). The outcome $1 + \text{int}(6 \times 0.60099\ldots) = 4$ is obtained if the random number $u = 0.60099\ldots$ is generated.

For the special case where each of the probabilities in a chance experiment is a multiple of a fraction $\frac{1}{r}$ for some integer r, you can use the ingenious *array method* to draw from the probability distribution. In this method you need only to generate one random integer from the integers 1 to r. To explain the method, consider a chance experiment with three possible outcomes O_1, O_2 and O_3 with respective probabilities $p_1 = 0.50$, $p_2 = 0.15$ and $p_3 = 0.35$. Then you form the array A[i] for $i = 1, \ldots, 100$ with A[1] $= \cdots =$ A[50] $= 1$, A[51] $= \cdots =$ A[65] $= 2$ and A[66] $= \cdots =$ A[100] $= 3$. You generate a random number u between 0 and 1. Next you calculate $k = 1 + \text{int}(100 \times u)$, being a random integer from $1, \ldots, 100$. Then A[m] gives you the random observation from the probability mass function. For example, suppose $u = 0.63044\ldots$ has been generated. Then, $m = 64$ with $A[64] = 2$. This gives the random outcome O_2.

5.2.4 Simulating a random permutation

Suppose you have 10 people and 10 labels numbered as 1 to 10. How to assign the labels at random such that each person gets assigned a different label? This can be done by making a random permutation of the integers $1, \ldots, 10$ and assigning the labels according to the random order in the permutation. An algorithm for generating a random permutation is useful for many probability problems. A simple and elegant algorithm can be given for generating a random permutation of $(1, 2, \ldots, n)$. The idea of the algorithm is first to randomly choose one of the integers $1, \ldots, n$ and to place that integer in position n. Next you randomly choose one of the remaining $n - 1$ integers and place it in position $n - 1$, etc.

Algorithm for random permutation

1. Initialize $t := n$ and $a[j] := j$ for $j = 1, \ldots, n$.

2. Generate a random number u between 0 and 1.

3. Set $k := 1 + \text{int}(t \times u)$ (random integer from the integers $1, \ldots, t$). Interchange the current values of $a[k]$ and $a[t]$.

4. $t := t - 1$. If $t > 1$, return to step 2; otherwise, stop with the random permutation $(a[1], \ldots, a[n])$.

As an illustration, the algorithm is used to construct a random permutation of the integers 1, 2, 3, and 4.

Iteration 1. $t := 4$. If the generated random number $u = 0.71397\ldots$, then $k = 1 + \text{int}(4 \times 0.71397\ldots) = 3$. Interchanging the elements of the positions $k = 3$ and $t = 4$ in $(1, 2, 3, 4)$ gives $(1, 2, 4, 3)$.
Iteration 2. $t := 3$. If the generated random number $u = 0.10514\ldots$, then $k = 1 + \text{int}(3 \times 0.10514\ldots) = 1$. Interchanging the elements of the positions $k = 1$ and $t = 3$ in $(1, 2, 4, 3)$ give $(4, 2, 1, 3)$.
Iteration 3. $t := 2$. If the generated random number $u = 0.05982\ldots$, then $k = 1 + \text{int}(2 \times 0.05982\ldots) = 1$. Interchanging the elements of the positions $k = 1$ and $t = 2$ in $(4, 2, 1, 3)$ gives $(2, 4, 1, 3)$.
Iteration 4. $t := 1$. The algorithm stops with the random permutation $(2, 4, 1, 3)$.

Verify yourselves that the algorithm carries literally over to the construction of a random permutation of a finite sequence of objects $(a_1, \ldots, a_n)$ in which some objects appear multiple times.

The algorithm can also be used to simulate a random subset of integers. For example, how to simulate a draw of the lotto 6/45 in which six distinct numbers are randomly drawn from the number 1 to 45? This can be done by using the algorithm with $n = 45$ and performing only the first 6 iterations until the positions $45, 44, \ldots, 40$ are filled. Then $a[45], \ldots, a[40]$ give the six numbers for the lottery draw.

5.2.5 Simulating a random point inside a circle

In subsection 5.2.2 you have seen how to generate a random point inside a rectangle. How do you generate a random point inside a circle? To do this, you face the complicating factor that the coordinates of a random point inside a circle cannot be generated independently of each other. Any point (x, y) inside a circle with radius r and the origin $(0, 0)$ as center must satisfy $x^2 + y^2 < r^2$. A tempting procedure is to generate first a random number $x = a$ from the interval $(-r, r)$ and to generate next a random number y from the interval $(-\sqrt{r^2 - a^2}, \sqrt{r^2 - a^2})$. This procedure, however, violates the requirement that the probability of the random point falling into a subregion should be the same for any two subregions having the same area.

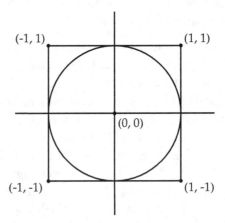

Figure 11: Simulating a random point inside the circle

A simple but powerful method to generate a random point inside the circle is the *hit-and-miss method*. The idea of this method is to take a rectangle that envelopes the bounded region and to generate random points inside the rectangle until a point is obtained that falls inside the circle. This simple approach can also be used to generate a random point inside a bounded region in the plane. As an illustration, take the unit circle with radius 1 and the origin $(0,0)$ as center. The circle is clamped into the square with the corner points $(-1,-1)$, $(1,-1)$, $(-1,1)$ and $(1,1)$, see Figure 11. A random point (x,y) inside this square is found by generating two random numbers u_1 and u_2 from $(0,1)$ and taking x and y as

$$x = -1 + 2 \times u_1 \quad \text{and} \quad y = -1 + 2 \times u_2.$$

Next you test whether
$$x^2 + y^2 < 1.$$

If this is the case, you have found a random point (x,y) inside the unit circle; otherwise, you repeat the procedure. On average, you have to generate $\frac{4}{\pi} = 1.273$ random points inside the square until you get a random point inside the circle. To see this, note that $\frac{4}{\pi}$ is the ratio of the area of the square and the area of the unit circle.

The idea of the hit-and-miss method is generally applicable. It is often used to find the area of a bounded region in the two-dimensional plane or a higher-dimensional space by enveloping the region by a rectangle or by a higher-dimensional cube. These areas are typically represented by multiple integrals. A classic example of the use of Monte Carlo simulation to compute such integrals goes back to the analysis of neutron diffusion problems in the atomic bomb research at the Los Alamos National Laboratory in 1944. The physicists Nicholas Metropolis and Stanislaw Ulam had to compute multiple integrals representing the volume of a 20-dimensional region and they devised the hit-and-miss method for that purpose.

5.3 Applications of computer simulation

In probability applications, simulation is a powerful tool for getting numerical answers to probability problems, which are otherwise too difficult for an analytical solution.[27] Several examples will be given to illustrate this. Simulation can also be used as a sanity check when you are not sure about the analytical solution.

5.3.1 Geometric probability problems

Geometric probability problems constitute a class of probability problems that often seem very simple but are sometimes very difficult to solve analytically. Take the problem of finding the expected value of the distance between two random points inside the unit square (sides with length 1) and the expected value of the distance between two

---

[27]In statistics, computer simulation is indispensable in the often used *bootstrapping method*. This method is used in situations that you have a representative random sample from a population and other samples from the population cannot be drawn. You want to learn from the data set how much a particular population statistic (e.g., the average or the median) is likely to vary. The idea of bootstrapping is to simulate many resamples from your data set. In each resample you randomly draw *with replacement* as many data points from your data set as there are in this set. This process allows you to calculate standard errors and confidence intervals for many types of sample statistics.

random points inside the unit circle (radius 1). The analytical derivation of these expected values requires advanced integral calculus and leads to the following exact results:

$$\frac{1}{15}\left[2 + \sqrt{2} + 5\ln(1 + \sqrt{2})\right] = 0.5214 \quad \text{and} \quad \frac{128}{45\pi} = 0.9054.$$

It is a piece of cake to estimate the expected values by computer simulation. How does the simulation program look like? Perform a very large number of simulation runs. In each simulation run two random points (x_1, y_1) and (x_2, y_2) are generated, using the method from subsection 5.2.2 for the unit square and using the method from subsection 5.2.5 for the unit circle. In each run the distance between the two points is calculated by Pythagoras as

$$\sqrt{(x_1 - x_2)^2 + (y_1 - y_2)^2}.$$

Then, the average of the distances found in the simulation runs is calculated. By the law of large numbers, the average gives an estimate for the expected value of the distance between two random points. Many simulation runs are needed to get accurate estimates. The question of how many runs should be done will be addressed in the next section. In one million simulation runs the estimates 0.5213 and 0.9053 were obtained for the expected values of the distances between two random point inside the unit square and between two random points inside the unit circle. For the interested reader Python computer programs are given in an appendix to this chapter. The computing times for the one million simulation runs are a matter of seconds on a fast computer.

5.3.2 Birthday problems

In the classic birthday problem the question is what the probability is that two or more people share a birthday in a randomly formed group of people. This problem has been analytically solved in Section 1.2. It was easy to find the answer. The problem becomes much more difficult when the question is what the probability is that two or more people have a birthday within one day from each other. This is the *almost-birthday problem*. However, the simulation program for the almost-birthday problem is just as simple as the simulation program for the birthday problem. An outline of the simulation programs is as follows. The starting point is a randomly formed group of m people (no twins), where each day is equally likely as birthday for any person. For ease, it is assumed that the year has 365 days (February 29 is excluded). For each of the two birthday problems, a very large number of simulation runs is performed. In each simulation run, m random integers $g_1, \ldots, g_m$ are generated, where the random integer g_i represents the birthday of the ith person. Each of these integers is randomly chosen from the integers $1, \ldots, 365$, using the simulation tool from subsection 5.2.3. In each simulation run for the classic birthday problem you test whether there are distinct indices i and j such that

$$|g_i - g_j| = 0,$$

while in each simulation run for the almost-birthday problem you test whether there are distinct indices i and j such that

$$|g_i - g_j| \leq 1 \text{ or } |g_i - g_j| = 364.$$

You find an estimate for the sought probability by dividing the number of simulation runs for which the test criterion is satisfied by the total number of runs. As you see, the simulation program for the almost-birthday problem is just as simple as that for the classic birthday problem.

5.3.3 Lottery problem

What is the probability of getting two or more consecutive numbers when six distinct numbers are randomly drawn from the numbers 1 to 45 in the lotto 6/45? The exact value of this probability is

$$1 - \frac{\binom{40}{6}}{\binom{45}{6}} = 0.5287,$$

but the argument to get this result is far from simple. However this probability, which is surprisingly large, can be quickly and easily obtained by computer simulation. In each simulation run you get the six lottery numbers by applying six iterations of the algorithm from subsection 5.2.4 and taking the six integers in the array elements $a[45], a[44], \ldots, a[40]$. Next you test whether there are two or more consecutive numbers among these six numbers. This is easily done by checking whether $a[i] - a[j]$ is 1 or -1 for some i and j with $40 \leq i, j \leq 45$. The desired probability is estimated by dividing the number of simulations runs for which the test criterion is satisfied by the total number of simulation runs. One million simulation runs resulted in the estimate 0.5289. The simulation program needs only a minor modification to simulate the probability of getting three or more consecutive numbers. Simulation leads to the estimate 0.056 for this probability.

5.3.4 The Mississippi problem

An amusing but very difficult combinatorial probability problem is the Mississippi problem. What is the probability that any two adjacent letters are different in a random permutation of the eleven letters of the word Mississippi? A simulation model can be constructed by identifying the letter m with the number 1, the letter i with the number 2, the letter s with the number 3, and the letter p with the number 4. In each simulation run a random permutation of the sequence $(1, 2, 3, 3, 2, 3, 3, 2, 4, 4, 2)$ is constructed by using an obvious modification of the permutation algorithm from subsection 5.2.4: the initialization of the algorithm now becomes $a[1] = 1$, $a[2] = 2$, ..., $a[10] = 4$, $a[11] = 2$. To test whether any two adjacent numbers are different in the resulting random permutation $(a[1], a[2], \ldots, a[11])$, you check whether $a[i + 1] - a[i] \neq 0$ for $i = 1, \ldots, 10$. The estimate 0.058 was obtained for the sought probability after 100 000 simulation runs.

It is never possible to achieve perfect accuracy through simulation. All you can measure is how likely the estimate is to be correct. This issue will be discussed in the next section. You will see that, if you want to achieve one more decimal digit of precision in the estimate, you have to increase the number of simulation runs with a factor of about one hundred. In other words, the probabilistic error bound decreases as the reciprocal square root of the number of simulation runs. The square root law is here at work.

5.4 Statistical analysis of simulation output

How many simulation runs should be made in order to get a desired level of accuracy in the estimate? When doing a simulation, it is important to have a probabilistic judgment about the accuracy of the point estimate. Such a judgment is provided by the concept of confidence interval. Suppose that you want to estimate the unknown probability p of a particular event E. To that end, n simulation runs are done. Let the indicator variable X_i be 1 if event E occurs in the ith simulation run and X_i be 0 otherwise. Then,

$$\hat{p} = \frac{1}{n} \sum_{i=1}^{n} X_i$$

is an estimator for the true value of p. The accuracy of this estimator is expressed by the 95% confidence interval

$$\left(\hat{p} - 1.96 \frac{\sqrt{\hat{p}(1-\hat{p})}}{\sqrt{n}}, \ \hat{p} + 1.96 \frac{\sqrt{\hat{p}(1-\hat{p})}}{\sqrt{n}} \right).$$

This confidence interval should be interpreted as follows: any such interval to be constructed by simulation will cover the true value of p with a probability of about 95% for large n. In other words, in about 95 out of 100 cases, the confidence interval covers the true value of p. Each simulation study gives an other confidence interval!

The effect of n on the term $\sqrt{\hat{p}(1-\hat{p})}$ fades away quickly if n gets larger. This means that the width of the confidence interval is nearly proportional to $1/\sqrt{n}$ for n sufficiently large. This conclusion leads to a practically important rule of thumb:

> **to reduce the width of a confidence interval by a factor of two, about four times as many observations are needed.**

This is a very useful rule for simulation purposes. Let's illustrate this with the almost-birthday problem with a group of 20 people. For the probability that two or more people will have their birthdays within one day of each other, a simulation with 25 000 runs results in the probability estimate of 0.8003 with (0.7953, 0.8052) as 95% confidence interval, whereas 100 000 simulation runs result in an estimate of 0.8054 with (0.8029, 0.8078) as 95% confidence interval. The confidence interval has indeed been narrowed by a factor of 2.

How can you construct a confidence interval if the simulation study is set up to estimate an unknown expected value of some random variable X rather than an unknown probability? Letting $X_1, \ldots, X_n$ represent the observations for X resulting from n independent simulation runs, then estimators for $\mu = E(X)$ and $\sigma = \sigma(X)$ are

$$\hat{\mu} = \frac{1}{n} \sum_{k=1}^{n} X_k \quad \text{and} \quad \hat{\sigma} = \sqrt{\frac{1}{n} \sum_{k=1}^{n} (X_k - \hat{\mu})^2}.$$

The 95% confidence interval for the unknown $\mu = E(X)$ is

$$\left(\hat{\mu} - 1.96 \frac{\hat{\sigma}}{\sqrt{n}}, \ \hat{\mu} + 1.96 \frac{\hat{\sigma}}{\sqrt{n}} \right).$$

The statistic $\hat{\sigma}/\sqrt{n}$, which is the estimated standard deviation of the *sample mean* $\hat{\mu}$, is usually called the *standard error* of the sample mean. Any 95% confidence interval to be constructed by simulation will cover the true value of μ with a probability of about 95% for n large. To illustrate, let the random variable X be the distance between two randomly chosen point in the unit square, see section 5.3.1. One million simulation runs resulted in $\hat{\mu} = 0.5213$ and $\hat{\sigma} = 0.2477$, which gives the confidence interval $(0.5209, 0.5218)$ for μ.

   To conclude this section, let's briefly sketch how the 95% confidence interval is obtained from the central limit theorem. By this theorem, $(\frac{1}{n} \sum_{i=1}^{n} X_i - \mu)/(\sigma/\sqrt{n})$ has approximately a standard normal distribution for n large. This result remains true when σ is replaced by its estimator $\hat{\sigma}$. The standard normal distribution has 95% of its mass between the percentiles $z_{0.025} = -1.96$ and $z_{0.975} = 1.96$. Therefore $P(-1.96 \le (\frac{1}{n} \sum_{i=1}^{n} X_i - \mu)/(\sigma/\sqrt{n}) \le 1.96) \approx 0.95$ for n large. This can be rewritten as

$$P\left(\frac{1}{n} \sum_{i=1}^{n} X_i - 1.96 \frac{\hat{\sigma}}{\sqrt{n}} \le \mu \le \frac{1}{n} \sum_{i=1}^{n} X_i + 1.96 \frac{\hat{\sigma}}{\sqrt{n}} \right) \approx 0.95.$$

Voila, the 95% confidence interval $(\hat{\mu} - 1.96\frac{\hat{\sigma}}{\sqrt{n}}, \hat{\mu} + 1.96\frac{\hat{\sigma}}{\sqrt{n}})$. The estimated standard deviation $\hat{\sigma}$ does not vary much if n gets larger and so the width of the confidential interval is proportional to $\frac{1}{\sqrt{n}}$ for larger values of n, implying that you need about four times as many simulation runs in order to reduce the width of the interval by a factor of two.

Simulation modeling problems

In each of the following modeling problems you are asked to set up a mathematical model for a simulation program for the problem in question. If you master a programming language, such as Python or R, it is fun to write a simulation program for some of the problems.

Problem 5.1. Set up a simulation model in order to estimate the probability that the equation $Ax^2 + Bx + C = 0$ has two real roots if A, B and C are randomly chosen numbers from $(-1, 1)$. Do the same if A, B and C are randomly chosen nonzero integers between $-1\,000$ and $1\,000$.

Problem 5.2. You draw random numbers from $(0, 1)$ until the sum of the picked numbers is larger than 1. Set up a simulation model to estimate the expected value of the number of picks needed.

Problem 5.3. You draw random numbers from $(0, 1)$ until you pick a random number that is smaller than the previous pick. Set up a simulation model in order to estimate the expected value of the number of picks needed.

Problem 5.4. You choose two random points on a stick. At these points you break the stick into three pieces. Set up a simulation model in order to estimate the probability that the largest piece is longer than the sum of the other two pieces.

Problem 5.5. You randomly pick two points on a stick. At these points the stick is broken into three pieces. Set up a simulation model in order to estimate the probability that a triangle can be formed with the pieces.

Problem 5.6. You randomly choose two points inside a circle. Set up a simulation model in order to estimate the probability that these two random points and the center of the circle form an obtuse triangle. Do the same for a sphere in which two random points are chosen.

Problem 5.7. Set up a simulation model in order to estimate the expected value of the area of the triangle that is formed by three randomly chosen points inside the unit square. Do the same for three randomly chosen points inside the unit circle. *Hint*: the area of a triangle with sides of lengths a, b and c is $\sqrt{s(s-a)(s-b)(s-c)}$, where $s = \frac{1}{2}(a+b+c)$.

Problem 5.8. Let $f(x)$ be a positive function on a finite interval (a, b) such that $0 \le f(x) \le M$ for $a \le x \le b$. How can you use simulation to estimate the integral $\int_a^b f(x)\,dx$?

Problem 5.9. Set up a simulation model in order to estimate the expected value of the distance between two randomly chosen points inside the unit cube. Do the same for the unit sphere.

Problem 5.10. A die is rolled until either each of the six possible outcomes has appeared or one of these outcomes has appeared six times. Set up a simulation model in order to estimate the probability that the first event will occur before the second event.

Problem 5.11. Seven students live in a same house. Set up a simulation model in order to estimate the probability that two or more of them have their birthdays within one week of each other.

Problem 5.12. The eight teams that have reached the quarter-finals of the Champions League soccer consist of two British teams, two German teams, two Italian teams and two Spanish teams. Set up a simulation model in order to estimate the probability that no two teams from the same country will be paired in the quarter-finals draw if the eight teams are paired randomly.

Problem 5.13. Sixteen teams remain in a soccer tournament. A drawing of lots will determine which eight matches will be played. Before the draw takes place, it is possible to place bets with book-makers about the results of the draw. Set up a simulation model in order to estimate the probability mass function of the number of correctly predicted matches.

Problem 5.14. Consider Problem 3.2 again. Set up a simulation model in order to estimate the probability of drawing the same pupil's name three or more times.

Problem 5.15. Consider Problem 2.40 again. Set up a model for simulating the probability mass function of the number of unpecked chickens.

Problem 5.16. Consider the Pólya urn model in which an urn initially contains r red and b black balls. At each time step, one ball is drawn randomly from the urn and put back in the urn along with an extra ball of the color drawn. The remarkable property of the Pólya urn model is that it has a random limit. Set up a simulation model in order to verify for the special case of $r = b = 1$ that the probability histogram of the fraction of red balls in the urn after a fixed number of n steps can be very well approximated by the uniform probability density on $(0, 1)$ for n large.

Problem 5.17. You play a game with 11 cards: ace, two, three,..., nine, ten, and joker. Each card counts for its face value with ace $= 1$ and joker $= 0$. The randomly ordered cards are turned over one by one until you decide to stop. Your score is the sum of the face values of the cards turned over as long as the joker has not shown up. If the joker appears, the game is over and your end score is zero. Consider the strategy of stopping when your score is 28 or more (this stopping rule maximizes the expected end score) and the strategy of stopping when your score is 15 or more. Set up a simulation model in order to estimate the probability of an end score of zero and the expected value of your end score for each of the two cases.

Problem 5.18. You repeatedly roll two fair dice until each of the 11 possible values of the sum of a roll with two dice has appeared.

Set up a simulation model in order to estimate the expected value of
the number of rolls needed.

Problem 5.19. Set up a simulation model in order to estimate the
probability of having at least two cards of the same face value (e.g.,
two nines or two queens) being next to each other in a randomly
shuffled standard deck of 52 cards.

Problem 5.20. Set up a simulation model in order to estimate the
probability of getting either five or more consecutive heads or five or
more consecutive tails, or both, in 25 tosses of a fair coin.[28]

Problem 5.21. A standard deck of 52 playing cards is thoroughly
shuffled after which the cards are turned over one by one. Set up a
simulation model in order to estimate the probability that either five
or more red cards or five or more black cards, or both, will appear
in a row.

Problem 5.22. You play the following coin tossing game. A fair
coin is tossed 20 times. You win if the sequence HHHH shows up
and not HHHT, you lose if HHHT shows up and not HHHH, and
otherwise you neither win nor lose. Set up a simulation model in
order to estimate your probability of winning and your probability
of losing.

Problem 5.23. Consider Problem 3.9 again. You now follow the
strategy of betting each time two seventh of your current bankroll.
Set up a simulation model in order to estimate the probability that
your end capital will be more than $100.

---

[28]Surprisingly long runs can occur in coin tossing. A rule of thumb says that the
probability mass function of the longest run of either heads or tails, or both, in n
tosses of a fair coin is strongly concentrated around $\log_2(\frac{1}{2}n) + 1$ for larger values
of n. A true story in this regard is the following. At the end of the 19th century,
the Le Monaco newspaper regularly published the results of roulette spins in the
casino of Monte Carlo. The famous statistician Karl Pearson (1857–1936) studied
these data to test his theories. He observed that red and black came up a similar
number of times, but also noticed that lengths of runs of either reds or blacks
were much shorter than he would have expected. What caused that? Well, it
turned out that the journalists at Le Monaco just made up the roulette results
at the bar of the casino. They didn't think anybody would notice.

Problem 5.24. In a class of 15 children, each child buys a small Santa Claus present for one of the other children. The gifts are numbered from 1 to n and each child knows the number of their own gift. Also, the children put their names on slips of papers. The 15 slips of paper and 15 cards with the numbers of the gifts are put in a box. Each child then draws at random a slip of paper and a card from the box. The child named on the slip gets the gift corresponding to the number on the card. Set up a simulation model in order to estimate the probability that there is at least one pair of children who exchange the gift they had bought given that none of the children has picked their own gift.

Problem 5.25. A Christmas party is held for 10 persons. Each person brings a gift to the party for a gift exchange. The gifts are numbered 1 to 10 and each person knows the number of their own gift. Cards with the numbers $1, \ldots, 10$ are put in a hat. The party goers consecutively pull a card out of the hat in order to determine the present he or she will receive. If a person pulls out a card corresponding to the number of their own gift, then the card is put back in the hat and that person draws another card. Set up a simulation model in order to estimate the probability that the last person gets stuck with his or her own gift. *Note*: imagine that each time the next person to take a card is chosen at random.

Problem 5.26. A famous TV show is The Tonight Show with Jimmy Fallon. In this show Jimmy plays the Egg Russian Roulette game with a guest of the show. The guest is always a celebrity from sports or film. The guest and Jimmy take turns picking an egg from a carton and smashing it on their heads. The carton contains a dozen eggs, four of which are raw and the rest are boiled. Neither Jimmy nor the guest knows which eggs are raw and which are boiled. The first person who has cracked two raw eggs on their head loses the game. The guest is the first to choose an egg. Set up a simulation model in order to estimate the probability that the guest will lose the game.

Problem 5.27. A retired gentleman considers participation in an investment fund. An adviser shows him that with a fixed yearly return of 14%, which was realized in the last few years, he could withdraw $15 098 from the fund at the end of each of the coming 20 years. This is music to the ears of the retired gentleman, and he decides to invest $100 000. However, the yearly return fluctuates. If the return was $r\%$ for the previous year, then for the coming year the return will remain at $r\%$ with a probability of 0.5, will change to $0.8r\%$ with a probability of 0.25, and will change to $1.2r\%$ with a probability of 0.25. Set up a simulation model for a probability histogram of the invested capital after 15 years when at the end of each of those 15 years $15 098 is withdrawn with the stipulation that the entire amount remaining in the fund is withdrawn if the remaining capital is less than $15 098.

Problem 5.28. A beer company brings a new beer with the brand name Babarras to the market and prints one of the letters of this brand name underneath each bottle cap. Each of the letters A, B, R, and S must be collected a certain number of times in order to get a nice beer glass. The quota for the letters A, B, R, and S are 3, 2, 2, and 1. These letters appear with probabilities 0.15, 0.10, 0.40, and 0.35, where the letters underneath the bottle caps are independent of each other. Set up a simulation in order to estimate the expected value and the probability mass function of the number of bottles that must be purchased in order to form the word Babarras.

Problem 5.29. An opaque bowl contains 11 envelopes in the colors red and blue. You are told that there are four envelopes of one color each containing $100 and seven empty envelopes of the other color, but you cannot see the envelopes in the bowl. The envelopes are taken out of the bowl, one by one and in random order. Each time an envelope has been taken out, you must decide whether or not to open this envelope. Once you have opened an envelope, you get the money in that envelope (if any) and the process stops. Your stopping rule is to open the envelope drawn as soon as four or more envelopes of each color have been taken out of the bowl. Set up a simulation model in order to estimate your probability of winning $100.

Appendix: Python programs for simulation

The two Python programs in this appendix are meant to give the reader not having programming experience some idea of how a computer program for simulation works. Python is a relatively simple programming language and is easy to learn since it requires a unique syntax that focuses on readability. The text after # is comment, which is not read by the program but is meant as explanation for the reader.

Python code for the distance between two random points inside unit square

```
import numpy as np # package for numerical computations
import numpy.random as rnd # package for random numbers
import numpy.linalg as la # package for linear algebra computations

def randompointinunitsquare(): # returns a random point in the unit square
    return rnd.rand(2)

def simulation(n): # this function generates n pairs of points (P, Q)
    delta = np.zeros(n) # allocates an array of length n
    for i in range(n): # n pairs of points
        P = randompointinunitsquare()
        Q = randompointinunitsquare()
        delta[i] = la.norm(P − Q) # distance between P and Q
    return delta

def main(): # the parameters are set and the output is printed
    n = 1000000 # number of simulation runs
    rnd.seed(15534) # seed 15534 starts random number generator
    delta = simulation(n) # output of the simulation() function
    print('average :', np.mean(delta)) # expected value of distance
    print('variance:', np.var(delta)) # variance of distance

main()
```

Python code for the distance between two random points
inside unit circle

```python
import numpy as np # package for numerical computations
import numpy.random as rnd # package for random numbers
import numpy.linalg as la # package for linear algebra computations

def randompointinunitcirle(): # returns a random point P in unit circle
    hit = 0
    while hit==0: # hit-and-miss method
        P = -1 + 2× rnd.rand(2) # random point in square
        hit = la.norm(P)< 1 # accept P if P in circle, otherwise repeat
    return P

def simulation(n): # this function generates n pairs of points (P,Q)
    delta = np.zeros(n) # allocates an array of length n
    for i in range(n): # n pairs of points
        P = randompointinunitcirle()
        Q = randompointinunitcirle()
        delta[i] = la.norm(P - Q) # distance between P and Q
    return delta

def main(): # the parameters are set and the output is printed
    n = 1000000 # number of simulation runs
    rnd.seed(15534) # seed 15534 starts random number generator
    delta = simulation(n) # output of the simulation() function
    print('average :', np.mean(delta)) # expected value of distance
    print('variance:', np.var(delta)) # variance of distance

main()
```

Chapter 6

A Primer on Markov Chains

This chapter introduces you to the very basic concepts and features of Markov chains. A Markov chain is basically a sequence of random variables evolving over time and having a weak form of dependency between them. This very useful model was developed in 1906 by Russian mathematician A.A. Markov (1856–1922). In a famous paper written in 1913, he used his probability model to analyze the frequencies at which vowels and consonants occur in Pushkin's novel "Eugene Onegin." Markov showed empirically that adjacent letters in Pushkin's novel are not independent but obey his theory of dependent random variables. The characteristic property of a Markov chain is that its memory goes back only to the most recent state. The Markov chain model is a very powerful probability model that is used today in countless applications in many different areas, such as voice recognition, DNA analysis, stock control, telecommunications and a host of others. The model is no exception to the rule that simple models are often the most useful models for analyzing practical problems.

6.1 Markov chain model

A Markov chain can be seen as a dynamic stochastic process that randomly moves from state to state with the property that only the current state is relevant for the next state. In other words, the memory of the process goes back only to the most recent state. A

picturesque illustration of this would show the image of a frog jump-
ing from lily pad to lily pad with appropriate transition probabilities
that depend only on the position of the last lily pad visited. In order to plug a specific problem into a Markov chain model, the state
variable(s) should be appropriately chosen in order to ensure the
characteristic memoryless property of the process. The basic steps
of the modeling approach are:

- Choosing the state variable(s) such that the current state summarizes everything about the past that is relevant to the future
  states.

- The specification of the one-step transition probabilities of moving from state to state in a single step.

Using the concept of state and choosing the state in an appropriate
way, surprisingly many probability problems can be solved within
the framework of a Markov chain. In this chapter it is assumed that
the set of states to be denoted by I is *finite*. This assumption is important. The theory of Markov chains involves quite some subtleties
when the state space is countably infinite.

The following notation is used for the one-step transition probabilities:

$$p_{ij} = \text{the probability of going from state } i \text{ to state } j \text{ in one step}$$

for $i, j \in I$. Note that the one-step probabilities must satisfy $p_{ij} \geq 0$
for all $i, j \in I$ and $\sum_{j \in I} p_{ij} = 1$ for all $i \in I$.

Example 6.1. A faulty digital video conferencing system has a
clustering error pattern. If a bit is received correctly, the probability
of receiving the next bit correctly is 0.99. This probability is 0.95 if
the last bit was received incorrectly. What is an appropriate Markov
chain model?

Solution. The choice of the state is obvious in this example. Let
state 0 mean that the last bit sent is not received correctly and state
1 mean that the last bit sent is received correctly. The sequence

of states is described by a Markov chain with one-step transition probabilities $p_{00} = 0.05$, $p_{01} = 0.95$, $p_{10} = 0.01$ and $p_{11} = 0.99$.

The choice of an appropriate state is more tricky in the next example. Putting yourself in the shoes of someone who must write a simulation program for the problem in question may be helpful in choosing the state variable(s).

Example 6.2. An absent-minded professor drives every morning from his home to the office and at the end of the day from the office to home. During the day, his driver's license is located at his home or at the office. If his driver's license is at his location of departure, he takes it with him with probability 0.5. What is an appropriate Markov chain model for this situation?

Solution. Your first thought might be to define two states 1 and 0, where state 1 describes the situation that the professor has his driver's license with him when driving his car and state 0 describes the situation that he does not have his driver's license with him when driving his car. However, these two states do not suffice for a Markov model: state 0 does not provide enough information to predict the state at the next drive (why?). In order to give the probability distribution of this next state, you need information about the current location of the driver's license of the professor. You get a Markov model by simply inserting this information into the state description. Therefore the following three states are defined:

- state 1 = the professor is driving his car and has his driver's license with him,

- state 2 = the professor is driving his car and his driver's license is at the point of departure,

- state 3 = the professor is driving his car and his driver's license is at his destination.

This state description has the Markovian property that the present state contains sufficient information for predicting future states. The

Markov chain model has state space $I = \{1, 2, 3\}$. What are the one-step transition probabilities p_{ij}? The only possible one-step transitions from state 1 are to the states 1 and 2 (verify!). A one-step transition from state 1 to state 1 occurs if the professor does not forget his license on the next trip and so $p_{11} = 0.5$. By a similar argument, $p_{12} = 0.5$. Obviously, $p_{13} = 0$. A one-step transition from state 2 is always to state 3, and so $p_{23} = 1$ and $p_{21} = p_{22} = 0$. The only possible transitions from state 3 are to the states 1 and 2. A one-step transition from state 3 to state 1 occurs if the professor takes his license with him on the next trip, and so $p_{31} = 0.5$. Similarly, $p_{32} = 0.5$. Obviously, $p_{33} = 0$. A *matrix* is the most useful way to display the one-step transition probabilities:

$$
\begin{array}{c|ccc}
\text{from}\backslash\text{to} & 1 & 2 & 3 \\
\hline
1 & 0.5 & 0.5 & 0 \\
2 & 0 & 0 & 1 \\
3 & 0.5 & 0.5 & 0
\end{array}.
$$

Time-dependent analysis of Markov chains

In Markov chains a key role is played by the n-step transition probabilities. For any $n = 1, 2, \ldots$, these probabilities are defined as

$$
p_{ij}^{(n)} = \text{the probability of going from state } i \text{ to state } j \text{ in } n \text{ steps}
$$

for all $i, j \in I$. Note that $p_{ij}^{(1)} = p_{ij}$. How to calculate the n-step transition probabilities? It will be seen that these probabilities can be calculated by matrix products. Many calculations for Markov chains can be boiled down to matrix calculations.

The so-called Chapman–Kolmogorov equations for calculating the n-step transition probabilities $p_{ij}^{(n)}$ are

$$
p_{ij}^{(n)} = \sum_{k \in I} p_{ik}^{(n-1)} p_{kj} \quad \text{for all } i, j \in I \text{ and } n = 2, 3, \ldots.
$$

This recurrence relation says that the probability of going from state i to state j in n steps is obtained by summing the probabilities of the mutually exclusive events of going from state i to some state k in the first $n - 1$ steps and then going from state k to state j in the nth step. A formal proof proceeds as follows. For initial state i, let A be the event that the state is j after n steps of the process and B_k be the event that the state is k after $n - 1$ steps. Then, by the law of conditional probability, $P(A) = \sum_{k \in I} P(A \mid B_k) P(B_k)$. Since $P(A) = p_{ij}^{(n)}$, $P(B_k) = p_{ik}^{(n-1)}$ and $P(A \mid B_k) = p_{kj}$, the Chapman–Kolmogorov equations follow.

Let's now verify that the n-step transition probabilities can be calculated by multiplying the matrix of one-step transition probabilities by itself n times. To do so, let $\mathbf{P}$ be the matrix with the p_{ij} as entries and $\mathbf{P}^{(m)}$ be the matrix with the $p_{ij}^{(m)}$ as entries. In matrix notation, the Chapman–Kolmogorov equations read as

$$\mathbf{P}^{(n)} = \mathbf{P}^{(n-1)} \times \mathbf{P} \quad \text{for } n = 2, 3, \ldots .$$

This gives $\mathbf{P}^{(2)} = \mathbf{P}^{(1)} \times \mathbf{P} = \mathbf{P} \times \mathbf{P}$. Next, you get $\mathbf{P}^{(3)} = \mathbf{P}^{(2)} \times \mathbf{P} = \mathbf{P} \times \mathbf{P} \times \mathbf{P}$. Continuing in this way, you see that $\mathbf{P}^{(n)}$ is given by the n-fold matrix product $\mathbf{P} \times \mathbf{P} \times \cdots \times \mathbf{P}$, shortly written as $\mathbf{P}^n$. Thus $\mathbf{P}^{(n)} = \mathbf{P}^n$, which verifies that

> $p_{ij}^{(n)}$ is the (i, j)th entry of the n-fold matrix product $\mathbf{P}^n$.

Example 6.3. On the Island of Hope the weather each day is classified as sunny, cloudy, or rainy. The next day's weather depends only on today's weather and not on the weather of the previous days. If the present day is sunny, the next day will be sunny, cloudy, or rainy with probabilities 0.70, 0.10, and 0.20. The transition probabilities for the weather are 0.50, 0.28, and 0.22 when the present day is cloudy and they are 0.40, 0.30, and 0.30 when the present day is rainy. What is the probability that it will be sunny three days from now if it is cloudy today? What is the probability distribution of the weather on a given day far away?

Solution. These questions can be answered by using a three-state Markov chain. Let's say that the weather is in state S if it is sunny, in state C if it is cloudy and in state R if it is rainy. The evolution of the weather is described by a Markov chain with state space $I = \{S, C, R\}$. The matrix $\mathbf{P}$ of the one-step transition probabilities of this Markov chain is given by

$$
\begin{array}{c}
\text{from}\backslash\text{to} \quad S \quad\quad C \quad\quad R \\
\begin{array}{c} S \\ C \\ R \end{array}
\begin{pmatrix}
0.70 & 0.10 & 0.20 \\
0.50 & 0.28 & 0.22 \\
0.40 & 0.30 & 0.30
\end{pmatrix}.
\end{array}
$$

To find the probability of having sunny weather three days from now, you need the matrix product $\mathbf{P}^3$:

$$
\mathbf{P}^3 = \begin{pmatrix}
0.61160 & 0.23248 & 0.15592 \\
0.59784 & 0.23731 & 0.16485 \\
0.59040 & 0.23992 & 0.16968
\end{pmatrix}.
$$

From this matrix you read off that the probability of having sunny weather three days from now is $p_{CS}^{(3)} = 0.59784$ if it is cloudy today.

What is the probability distribution of the weather on a day far away? Intuitively, you expect that this probability distribution does not depend on the present state of the weather. This is indeed the case. Trying several values of n, it was found after 9 matrix multiplication that the elements of the matrix $\mathbf{P}^9$ agree row-to-row to four decimal places. That is,

$$
\mathbf{P}^{(n)} = \begin{pmatrix}
0.60497 & 0.23481 & 0.16022 \\
0.60497 & 0.23481 & 0.16022 \\
0.60497 & 0.23481 & 0.16022
\end{pmatrix} \quad \text{for all } n \geq 9.
$$

Thus the weather on a day far away will be sunny, cloudy, or rainy with respective probabilities of about 60.5%, 23.5% and 16.0%, regardless of the present weather. It is intuitively obvious that these probabilities also give long-run proportions of time that the weather will be sunny, cloudy, or rainy, respectively.

Many probability problems, which are seemingly unrelated to Markov chains, can be modeled as a Markov chain with the help of a little imagination. This is illustrated with the next example. This example nicely shows that the line of thinking through the concepts of state and state transition is very useful to analyze the problem (and many other problems in applied probability!).

Example 6.4. Six fair dice will be simultaneously rolled. What is the probability mass function of the number of different outcomes $1, 2, \ldots, 6$ that will show up?

Solution. In solving a probability problem, it is often helpful to recognize when two problems are equivalent even if they sound different. To put the dice problem into the framework of a Markov chain, consider the equivalent problem of repeatedly rolling a single die six times (the reformulated problem is an instance of the coupon collector's problem). For the reformulated problem, define the state of the process as the number of different face values seen so far. The evolution of the state is described by a Markov chain with state space $I = \{0, 1, \ldots, 6\}$, where state 0 is the initial state. The one-step transition probabilities are given by (verify!):

$$p_{01} = 1, \ p_{ii} = \frac{i}{6} \ \text{ and } \ p_{i,i+1} = 1 - \frac{i}{6} \ \text{ for } 1 \le i \le 5,$$
$$p_{66} = 1, \ \text{ and the other } p_{ij} = 0.$$

The desired probability mass function is obtained by calculating the probability $p_{0k}^{(6)}$ for $k = 1, \ldots, 6$, which gives the probability of getting exactly k different face values when rolling a single die six times. Multiplying the matrix $\mathbf{P}$ of one-step transition probabilities six times by itself results in the matrix $\mathbf{P}^{(6)}$:

$$\begin{pmatrix}
0 & 0.0001 & 0.0199 & 0.2315 & 0.5015 & 0.2315 & 0.0154 \\
0 & 0.0000 & 0.0068 & 0.1290 & 0.4501 & 0.3601 & 0.0540 \\
0 & 0 & 0.0014 & 0.0570 & 0.3475 & 0.4681 & 0.1260 \\
0 & 0 & 0 & 0.0156 & 0.2165 & 0.5248 & 0.2431 \\
0 & 0 & 0 & 0 & 0.0878 & 0.4942 & 0.4180 \\
0 & 0 & 0 & 0 & 0 & 0.3349 & 0.6651 \\
0 & 0 & 0 & 0 & 0 & 0 & 1
\end{pmatrix}.$$

You read off from the first row of the matrix $\mathbf{P}^{(6)}$ that $p_{01}^{(6)} = 0.0001$, $p_{02}^{(6)} = 0.0199$, $p_{03}^{(6)} = 0.2315$, $p_{04}^{(6)} = 0.5015$, $p_{05}^{(6)} = 0.2315$, $p_{06}^{(6)} =$

0.0154. Note that $1 - p_{06}^{(6)} = 0.9846$ gives the probability that more than 6 rolls of a die will be needed to get all six possible outcomes.

Problem 6.1. Consider Example 6.2 again. It is Wednesday evening and the professor is driving to home, unaware of the fact that there will be traffic control on the roadway to his house coming Friday evening. What is the probability that the professor will be fined for not having his license with him given that he left his license at the university on Wednesday evening? (answer: 0.625)

Problem 6.2. Every day, it is either sunny or rainy on Rainbow Island. The weather for the next day depends only on today's weather and yesterday's weather. The probability that it will be sunny tomorrow is 0.9 if the last two days were sunny, is 0.45 if the last two days were rainy, is 0.7 if today's weather is sunny and yesterday's weather was rainy, and is 0.5 if today's weather is rainy and yesterday's weather was sunny. Define a Markov chain that describes the weather on Rainbow Island and specify the one-step transition probabilities. What is the probability of having sunny weather five days from now if it rained today and yesterday? (answer: 0.7440)

Problem 6.3. An airport bus deposits 25 passengers at 7 stops. Each passenger is as likely to get off at any stop as at any other, and the passengers act independently of one another. The bus makes a stop only if someone wants to get off. Use Markov chain analysis to calculate the probability mass function of the number of bus stops. (answer: (0, 0.0000, 0.0000, 0.0000, 0.0046, 0.1392, 0.8562))

Problem 6.4. Consider Example 6.3 again. Use indicator random variables to calculate the expected value of the number of sunny days in the coming seven days given that is cloudy today. (answer: 4.049)

6.2 Absorbing Markov chains

A powerful trick in Markov chain analysis is to use one or more absorbing states. A state i of a Markov chain is said to be *absorbing* if $p_{ii} = 1$, that is, once the process enters an absorbing state i, it always stays there. Absorbing Markov chains are very useful to analyze success runs. This is illustrated with the following two examples.

Example 6.3 (continued). What is probability there will be three or more consecutive days with sunny weather in the coming 14 days given that it is rainy today?

Solution. Augment the three states in the Markov chain model from Example 6.3 with two additional states SS and SSS, where state SS means that it was sunny the last two days and state SSS means that it was sunny the last three days. State SSS is taken as an absorbing state. The matrix $\mathbf{P}$ of one-step transition probabilities now becomes

from\to	S	C	R	SS	SSS
S	0	0.10	0.20	0.70	0
C	0.50	0.28	0.22	0	0
R	0.40	0.30	0.30	0	0
SS	0	0.10	0.20	0	0.70
SSS	0	0	0	0	1

Since the process stays in state SSS once it is there, three consecutive days with sunny weather occur somewhere in the coming 14 days if and only if the process is in state SSS 14 days hence. Thus $p_{R,SSS}^{(14)}$ gives the probability that there will three or more consecutive days with sunny weather in the coming 14 days given that it is rainy today. The matrix $\mathbf{P}^{(14)}$ is obtained by multiplying the matrix $\mathbf{P}$ by itself 14 times and is given by

$$\mathbf{P}^{(14)} = \begin{pmatrix} 0.0214 & 0.0186 & 0.0219 & 0.0178 & 0.9203 \\ 0.0320 & 0.0278 & 0.0328 & 0.0265 & 0.8809 \\ 0.0334 & 0.0290 & 0.0342 & 0.0277 & 0.8757 \\ 0.0117 & 0.0102 & 0.0120 & 0.0097 & 0.9564 \\ 0 & 0 & 0 & 0 & 1 \end{pmatrix}.$$

You read off from row 3 that the desired probability equals 0.8757.

Example 6.5. What is the probability of getting either five or more consecutive heads or five or more consecutive tails, or both, in 25 tosses of a fair coin?

Solution. Use a Markov chain with six states $0, 1, \ldots, 5$. State 0 corresponds to the start of the coin-tossing experiment. For $i = 1, \ldots, 5$, state i means that the last i tosses constitute a run of length i, where a run consists of only heads or only tails. State 5 is taken as an absorbing state. The one-step transition probabilities of the Markov chain are $p_{01} = 1$, $p_{i,i+1} = p_{i1} = 0.5$ for $1 \leq i \leq 4$, $p_{55} = 1$ and the other $p_{ij} = 0$. The probability of getting either five or more consecutive heads or five or more consecutive tails, or both, in 25 tosses is $p_{05}^{(25)}$. Calculating

$$
\mathbf{P}^{25} = \begin{pmatrix}
0 & 0.2336 & 0.1212 & 0.0629 & 0.0326 & 0.5496 \\
0 & 0.2252 & 0.1168 & 0.0606 & 0.0314 & 0.5659 \\
0 & 0.2089 & 0.1084 & 0.0562 & 0.0292 & 0.5974 \\
0 & 0.1774 & 0.0920 & 0.0478 & 0.0248 & 0.6580 \\
0 & 0.1168 & 0.0606 & 0.0314 & 0.0163 & 0.7748 \\
0 & 0 & 0 & 0 & 0 & 1
\end{pmatrix},
$$

you find that $p_{05}^{(25)} = 0.5496$. A remarkably high probability! Most people grossly underestimate the lengths of longest runs.

Problem 6.5. What is the probability of getting five or more consecutive heads during 25 tosses of a fair coin? (answer: 0.3116)

Problem 6.6. What is the probability of getting the run 123456 during 500 rolls of a fair die? (answer: 0.0106)

Problem 6.7. What is the probability of getting two consecutive totals of 7 before a total of 12 when repeatedly rolling two dice? (answer: 0.4615)

Problem 6.8. John and Pete play a game using a plain die with red stickers on four faces and blue stickers on the other two. The die is rolled until either the color red has shown up four times in a row (win for John) or the color blue two times in a row (win for Pete) or ten rolls have been done. What are the win probabilities of John and Pete? (answer: 0.4325 and 0.4459)

Problem 6.9. You toss a fair coin until HTH or HHT appears. What is the probability that HHT appears first? (answer: $\frac{2}{3}$)

Problem 6.10. A fair coin is tossed 20 times. What is the probability that the sequence HHHH shows up and what is the probability that the sequence HHHT shows up? (answer: 0.4780 and 0.7462)

Problem 6.11. You are fighting a dragon with two heads. Each time you swing at the dragon with your sword, there is a 75% chance of knocking off one head and a 50% chance of missing. If you miss, either one additional head or two additional heads will grow immediately before you can swing again at the dragon. The probability of one additional head is 0.7 and of two additional heads is 0.3. You win if you have knocked off all of the dragon's heads, but you must run for your life if the dragon has five or more heads. Use Markov chain analysis to calculate your chance of winning. (answer: 0.5210)

Problem 6.12. In the dice game of Pig, you repeatedly roll a single die. Upon rolling a 1, your turn is over and you get a score zero. Otherwise, you can stop whenever you want and then your score is the total number of points rolled. Under the hold-at-20 rule you stop when you have rolled 20 points or more.[29] Use Markov chain analysis to get the probability mass function of your end score under the hold-at-20 rule. (answer: (0.6245, 0.0997, 0.0950, 0.0742, 0.0542, 0.0352, 0.0172) on (0, 20, 21, 22, 23, 24, 25) with E(end score) = 8.14)

Problem 6.13. Joe Dalton desperately wants to raise his bankroll of \$600 to \$1 000 in order to pay his debts before midnight. He enters a casino to play European roulette. He decides to bet on red each time using bold play, that is, Joe bets either his entire bankroll or the amount needed to reach the target bankroll, whichever is smaller. Thus the stake is \$200 if his bankroll is \$200 or \$800 and the stake is \$400 if his bankroll is \$400 or \$600. What is the probability that Joe will reach his goal? (answer: 0.5819)

Problem 6.14. Consider Problem 5.26 from Chapter 5 again. Use Markov chain analysis to compute the probability that the guest will lose the game. (answer: $\frac{5}{9}$)

---

[29]The rationale behind this stopping rule is as follows. Suppose your current score is x points and you decide for one other roll of the die. Then, the expected value of the change of your score is $\sum_{k=2}^{6} \frac{1}{6} \times k - \frac{1}{6} \times x = \frac{20}{6} - \frac{x}{6}$, which is non-positive for $x \geq 20$. This is the principle of the one-stage-look-ahead rule.

6.3 The gambler's ruin problem

A nice illustration of an absorbing Markov chain is the gambler's ruin problem that goes back to Christiaan Huygens (1629–1695) and Blaise Pascal (1623–1662). This random walk problem will be used to demonstrate that the absorption probabilities can also be calculated by solving a system of linear equations instead of taking matrix products. The method of linear equations can also be used to calculate the expected time until absorption. The gambler's ruin problem is as follows. Two players A and B with initial bankrolls of a dollars and b dollars play a game until one of the players is bankrupt. In a play of the game player a wins one dollar from player B with probability p and loses one dollar to player B with probability $q = 1 - p$. The successive plays of the game are independent of each other. What is the probability $P(a, b)$ that player A is the ultimate winner and what is the expected value $E(a, b)$ of the number of plays until one of the players goes broke?

The quantities $P(a, b)$ and $E(a, b)$ can be found by using an absorbing Markov chain. The Markov chain has the states $0, 1, \ldots, a+b$, where state i means that the current bankroll of player A is i dollars (and that of player B is $a + b - i$ dollars). The states 0 and $a + b$ are taken as absorbing states. The one-step transition probabilities are $p_{i,i+1} = p$, $p_{i,i-1} = q$ and the other $p_{ij} = 0$ for $i = 1, 2, \ldots, a + b - 1$. The probability $P(a, b)$ that player A will be the ultimate winner can be found by calculating the n-step transition probability $p_{a,a+b}^{(n)}$ for sufficiently large values of n. However, a more elegant approach is as follows. For $i = 0, 1, \ldots, a + b$, define f_i as the probability that the Markov chain will be ultimately absorbed in state $a+b$ when the starting state is i. By definition, $f_0 = 0$ and $f_{a+b} = 1$. The other f_i can be found by solving the linear equations

$$f_i = pf_{i+1} + qf_{i-1} \quad \text{for } i = 1, 2, \ldots, a + b - 1.$$

In particular, f_a gives the desired probability $P(a, b)$. The equation for f_i is easily explained from the law of conditional probability: the term pf_{i+1} accounts for the case of a win of player A in state i and the term qf_{i-1} for a win of player B in state i. Since the equations for the f_i are so-called linear difference equations, they can be explicitly solved. The details are omitted and we state only the famous *gambler's ruin formula* for $f_a = P(a, b)$:

$$P(a, b) = \begin{cases} \frac{1-(q/p)^a}{1-(q/p)^{a+b}} & \text{if } p \neq q \\ \frac{a}{a+b} & \text{if } p = q. \end{cases}$$

To get $E(a, b)$, define e_i as the expected value of the number of transitions of the Markov chain needed to reach either state 0 or state $a + b$ when starting from state i. By definition, $e_0 = e_{a+b} = 0$. The other e_i can be found by solving the linear equations

$$e_i = 1 + pe_{i+1} + qe_{i-1} \quad \text{for } i = 1, 2, \ldots, a+b-1.$$

Here we use the law of conditional expectation: the number of plays from starting state i is distributed with probability p as one plus the number of plays from starting state $i + 1$ and with probability q as one plus the number of plays from starting state $i + 1$. Thus the term 1 counts for the first play and $pe_{i+1} + qe_{i-1}$ for the expected number of additional plays. The linear equations for the e_i can also be explicitly solved. In particular, $e_a = E(a, b)$ is given by

$$E(a, b) = \begin{cases} \frac{a}{q-p} - \frac{a+b}{q-p} \frac{1-(q/p)^a}{1-(q/p)^{a+b}} & \text{if } p \neq q \\ ab & \text{if } p = q. \end{cases}$$

To illustrate the gambler's ruin formula, suppose you go to the casino of Monte Carlo with 100 euro, and your goal is to double it. You opt to play European roulette, betting each time on red. You will double your stake with probability $p = \frac{18}{37}$ and you will lose it with probability $q = \frac{19}{37}$. So, if you stake \$5 each time ($a = b = 20$), or \$10 ($a = b = 10$) or \$25 ($a = b = 4$), or \$50 ($a = b = 2$), the probability of reaching your goal will have the respective values of 0.2533, 0.3680, 0.4461 and 0.4730. The expected number of bets has the respective values 365.19, 97.66, 15.94 and 4.00. You see that if your only goal is to achieve the highest probability of winning a pre-determined target amount at the casino, it is better to bet large rather than small sums.

6.4 Long-run behavior of Markov chains

Let's consider a Markov chain with a finite set of states I (the assumption of a finite state space is important in the steady-state analysis of Markov chains). The process starts in one of the states and moves successively from one state to another, where the probability of moving from state i to state j in one step is denoted by p_{ij}. What about the probability distribution of the state after many, many transitions? Does the effect of the starting state ultimately fade away? In order to answer these questions, two conditions are introduced.

Condition C_1. The Markov chain has no two or more disjoint closed sets of states, where a set C is said to be closed if $p_{ij} = 0$ for $i \in C$ and $j \notin C$.

Condition C_2. The set of states cannot be split into multiple disjoint sets $S_1, \ldots, S_d$ with $d \geq 2$ such that a one-step transition from a state in S_k is always to a state in S_{k+1}, where $S_{d+1} = S_1$.

The condition C_1 is satisfied in nearly any application, but this is not the case for condition C_2. The condition C_2 rules out periodicity in the state transitions such as is the case in the three-state Markov chain with $p_{12} = p_{13} = 0.5$, $p_{21} = p_{31} = 1$, and $p_{ij} = 0$ otherwise. In this Markov chain, $p_{11}^{(n)}$ is alternately 0 or 1 and so $\lim_{n \to \infty} p_{11}^{(n)}$ does not exist (neither any of the $p_{ij}^{(n)}$ has a limit as n tends to infinity).

Under the conditions C_1 and C_2, the limiting probability (or equilibrium probability)

$$\pi_j = \lim_{n \to \infty} p_{ij}^{(n)}$$

exists for all $i, j \in I$ and is independent of the starting state i. The π_j can be calculated as the unique solution to the balance equations

$$\pi_j = \sum_{k \in I} p_{kj} \pi_k \quad \text{for } j \in I$$

together with the normalization equation $\sum_{j \in I} \pi_j = 1$. The balance equations can be easily explained from the Chapman-Kolmogorov equations $p_{ij}^{(n)} = \sum_{k \in I} p_{ik}^{(n-1)} p_{kj}$. Letting $n \to \infty$ in both sides of the Chapman–Kolmogorov equations and interchanging limit and summation (justified by the finiteness of I), you get the balance equations. The reader is asked to take for granted that the limits π_j exist and are uniquely determined by the set of linear equations.

As an illustration, consider Example 6.3 again. The balance equations together with the normalization equation are then given by

$$\pi_S = 0.70\pi_S + 0.50\pi_C + 0.40\pi_R$$
$$\pi_C = 0.10\pi_S + 0.28\pi_C + 0.30\pi_R$$
$$\pi_R = 0.20\pi_S + 0.22\pi_C + 0.30\pi_R$$
$$\pi_S + \pi_C + \pi_R = 1.$$

In solving these equations, you are allowed to delete one of the balance equations in order to get a square system of linear equations. Solving the last three equations gives $\pi_S = 0.60497$, $\pi_C = 0.23481$, and $\pi_R = 0.16022$. The same numerical answers as were obtained in Example 6.3 by calculating $\mathbf{P}^n$ for large n.

The probability π_j can be interpreted as the probability of finding the Markov chain in state j many, many transitions later, whatever the current state is. This interpretation is obvious. A second interpretation of π_j is that of a long run average: the proportion of transitions to state j will be π_j when averaging over many, many transitions. This second interpretation remains valid when the non-periodicity Condition C_2 is not satisfied, in which case it can be shown that

$$\pi_j = \lim_{n \to \infty} \frac{1}{n} \sum_{k=1}^{n} p_{ij}^{(k)} \quad \text{for all } j \in I,$$

independently of the starting state i. Again, the π_j are the unique solution to the balance equations $\pi_j = \sum_{k \in I} p_{kj} \pi_k$ for $j \in I$ together with the normalization equation $\sum_{j \in I} \pi_j = 1$. To illustrate this, consider again the example of a three-state Markov chain with one-step transition probabilities $p_{12} = p_{13} = 0.5$, $p_{21} = p_{31} = 1$, and $p_{ij} = 0$ otherwise. This particular Markov chain is periodic and $\lim_{n \to \infty} p_{ij}^{(n)}$ do not exist, e.g., $p_{11}^{(n)} = p_{22}^{(n)} = p_{33}^{(n)} = 0$ for n odd, whereas $p_{11}^{(n)} = 1$, $p_{22}^{(n)} = p_{33}^{(n)} = 0.5$ for n even. You can directly see that the long-run proportion of transitions to state j has the values $\frac{1}{2}$, $\frac{1}{4}$ and $\frac{1}{4}$ for $j = 1$, 2 and 3. Indeed, the balance equations $\pi_1 = \pi_2 + \pi_3$, $\pi_2 = 0.5\pi_1$ and $\pi_3 = 0.5\pi_1$ together with $\pi_1 + \pi_2 + \pi_3 = 1$ have the unique solution $\pi_1 = \frac{1}{2}$, $\pi_2 = \frac{1}{4}$ and $\pi_3 = \frac{1}{4}$.

Page-ranking algorithm

The theory of Markov chains has many applications. The most famous application is Markov's own analysis of the frequencies at which vowels and consonants occur in Pushkin's novel "Eugene Onegin," see also Problem 6.15 below.

An important application of more recent date is the application of Markov chains to the ranking of web pages. The page-ranking algorithm is one of the methods Google uses to determine a page's relevance or importance. Suppose that you have n interlinked web pages. Let n_j be the number of outgoing links on page j. It is assumed that $n_j > 0$ for all j. Let α be a given number with $0 < \alpha < 1$. Imagine that a random surfer jumps from his current page by choosing with probability α a random page amongst those that are linked from the current page, and by choosing with probability $1 - \alpha$ a completely random page. Hence the random surfer jumps around the web from page to page according to a Markov chain with the one-step transition probabilities

$$p_{jk} = \alpha\, r_{jk} + (1 - \alpha)\, \frac{1}{n} \quad \text{for } j, k = 1, \ldots, n,$$

where $r_{jk} = \frac{1}{n_j}$ if page k is linked from page j and $r_{jk} = 0$ otherwise. The parameter α was originally set to 0.85. The inclusion of the term $(1 - \alpha)/n$ can be justified by assuming that the random surfer occasionally gets bored and then randomly jumps to any page on the web. Since the probability of such a jump is rather small, it is reasonable that it does not influence the ranking very much. By the term $(1 - \alpha)/n$ in the p_{jk}, the Markov chain has no two or more disjoint closed sets and is aperiodic. Thus the Markov chain has a unique equilibrium distribution $\{\pi_j\}$. These probabilities can be estimated by multiplying the matrix $\mathbf{P}$ of one-step transition probabilities by itself repeatedly. Because of the constant $(1 - \alpha)/n$ in the matrix $\mathbf{P}$, things mix better up so that the n-fold matrix product $\mathbf{P}^n$ converges very quickly to its limit. The equilibrium probability π_j gives us the long-run proportion of time that the random surfer will spend on page j. If $\pi_j > \pi_k$, then page j is more important than page k and should be ranked higher.

Problem 6.15. In a famous paper written in 1913, Andrey Markov analyzed an unbroken sequence of 20 000 letters from the poem Eugene Onegin. He found that the probability of a vowel following a vowel is 0.128, and that the probability of a vowel following a consonant is 0.663. Use a Markov chain to estimate the percentages of vowels and consonants in the novel. (answer: 43.2% and 56.8%)

Problem 6.16. What is for Example 6.1 the long-run fraction of bits that are incorrectly received? (answer: $\frac{1}{96}$)

Problem 6.17. In a certain town, there are four entertainment venues. Both Linda and Bob are visiting every weekend one of these venues, independently of each other. Each of them visits the venue of the week before with probability 0.4 and chooses otherwise at random one of the other three venues. What is the long-run fraction of weekends that Linda and Bob visit a same venue? (answer: $\frac{1}{4}$).

Problem 6.18. Consider Problem 6.2 again. What is the long-run fraction of sunny days? (answer: 0.7912) The entrepreneur Jerry Wood has a pub on the island. On every sunny day, his sales are $N(\mu_1, \sigma_1^2)$ distributed with $\mu_1 =\$1\,000$ and $\sigma_1 = \$200$, while on rainy days his sales are $N(\mu_2, \sigma_2^2)$ distributed with $\mu_2 = \$500$ and $\sigma_2 = \$75$. What is the long-run average sales per day? (answer: \$895.60)

6.5 Markov chain Monte Carlo simulation[30]

This section gives a first introduction to Markov chain Monte Carlo (MCMC). This method can be used to tackle computational problems that arise among others in Bayesian inference. Let S be a very large but finite set on which a probability mass function $\pi(s)$ is given that is only known up to a multiplicative constant. It is not feasible

---

[30]This section contains advanced material. This material is taken from Henk Tijms, *Understanding Probability*, 3rd ed., Cambridge University Press, 2012.

to compute the constant directly. How to calculate $\sum_{s \in S} h(s)\pi(s)$ for a given function $h(s)$? The idea is to construct a Markov chain that has $\pi(s)$ as its equilibrium distribution and to simulate a sequence $s_1, s_2, \ldots, s_m$ of successive states of this Markov chain for large m. Then $\sum_{s \in S} h(s)\pi(s)$ can be estimated by $\frac{1}{m} \sum_{k=1}^{m} h(s_k)$, as will be shown below.

To explain the MCMC approach, the concept of *detailed balance* (or *reversibility*) is required. A probability mass function $\{a_j, j \in S\}$ on a finite set S is said to satisfy detailed balance equations with respect to an irreducible Markov chain with $\mathbf{P} = (p_{ij})$ as matrix of one-step transition probabilities if

$$\boxed{a_j p_{jk} = a_k p_{kj} \quad \text{for all } j, k \in S.}$$

A Markov matrix $\mathbf{P}$ is called *irreducible* if any state is reachable from any other state, that is, for any $i, j \in S$, there is some $n \geq 1$ such that $p_{ij}^{(n)} > 0$. Under the reversibility condition, the a_j are the unique equilibrium distribution of the Markov chain. To prove this, sum both sides of $a_j p_{jk} = a_k p_{kj}$ over $k \in S$. Together with $\sum_{k \in S} p_{jk} = 1$, this gives $a_j = \sum_{k \in S} a_k p_{kj}$ for all $j \in S$. These equations are the balance equations of the Markov chain with transition matrix $\mathbf{P}$.

Metropolis–Hastings algorithm

The Metropolis–Hastings algorithm is an example of a Markov chain Monte Carlo method. The algorithm will be first explained for the case of a discrete probability distribution, but the basic idea of the algorithm can be directly generalized to the case of a continuous probability distribution.

Let S be a very large but finite set of states on which a probability mass function $\{\pi(s), s \in S\}$ is given, where $\pi(s) > 0$ for all s and the $\pi(s)$ are only known up to a multiplicative constant. The Metropolis–Hastings algorithm generates a sequence of states $(s_1, s_2, \ldots)$ from a Markov chain that has $\{\pi(s), s \in S\}$ as its unique equilibrium distribution. To that end, the algorithm uses a candidate-transition function $q(t \mid s)$ (for clarity of presentation, the notation $q(t \mid s)$ is used rather than p_{st}). This function is to be interpreted as saying that when the current state is s the candidate for the next state is

t with probability $q(t \mid s)$. Thus you first choose, for each $s \in S$, a probability mass function $\{q(t \mid s), t \in S\}$. These functions must be chosen in such a way that the Markov matrix with the $q(t \mid s)$ as one-step transition probabilities is irreducible. The idea is to adjust these transition probabilities in such a way that the resulting Markov chain has $\{\pi(s), s \in S\}$ as unique equilibrium distribution. The reversibility equations are the key to this idea. If the candidate-transition function $q(t \mid s)$ already satisfies the detailed balance equations

$$\pi(s)q(t \mid s) = \pi(t)q(s \mid t) \quad \text{for all } s, t \in S,$$

you are done: the Markov chain with the $q(t \mid s)$ as one-step transition probabilities is reversible and has the $\pi(s)$ as its unique equilibrium distribution. What should you do when the detailed balance equations are not fully satisfied? The answer is to modify the transition probabilities by rejecting certain transitions. To work out this idea, fix two states s and t for which the detailed balance equation is not satisfied. It is no restriction to assume that $\pi(s)q(t \mid s) > \pi(t)q(s \mid t)$. Otherwise, reverse the roles of s and t. If $\pi(s)q(t \mid s) > \pi(t)q(s \mid t)$, then, loosely speaking, the process moves from s to t too often. How could you restore this? A simple trick to reduce the number of transitions from s to t is to use an acceptance probability $\alpha(t \mid s)$: the process is allowed to make the transition from s to t with probability $\alpha(t \mid s)$ and otherwise the process stays in the current state s. The question remains how to choose $\alpha(t \mid s)$. The choice of $\alpha(t \mid s)$ is determined by the requirement

$$\pi(s)[q(t \mid s)\alpha(t \mid s)] = \pi(t)[q(s \mid t)\alpha(s \mid t)].$$

Taking $\alpha(s \mid t) = 1$ for transitions from t to s, you get

$$\alpha(t \mid s) = \frac{\pi(t)q(s \mid t)}{\pi(s)q(t \mid s)}.$$

Therefore, for any $s, t \in S$, the acceptance probability is defined by

$$\boxed{\alpha(t \mid s) = \min\left[\frac{\pi(t)q(s \mid t)}{\pi(s)q(t \mid s)}, 1\right].}$$

Finally, the one-step transition probabilities to be used in the algorithm are defined by

$$q_{MH}(t \mid s) = \begin{cases} q(t \mid s)\alpha(t \mid s) & \text{for } t \neq s \\ 1 - \sum_{t \neq s} q(t \mid s)\alpha(t \mid s) & \text{for } t = s. \end{cases}$$

The Markov chain with these one-step transition probabilities satisfies the detailed balance equations $\pi(s)q_{MH}(t \mid s) = \pi(t)q_{MH}(s \mid t)$ for all s, t. Therefore this Markov chain has $\{\pi(s), s \in S\}$ as its unique equilibrium distribution. It is important to note that for the construction of the Markov chain it suffices to know the π_s up to a multiplicative constant because the acceptance probabilities involve only the ratio's $\pi(s)/\pi(t)$.

Summarizing, the Markov chain operates as follows. If the current state is s, a candidate state t is generated from the probability mass function $\{q(t \mid s), t \in S\}$. If $t \neq s$, then state t is accepted with probability $\alpha(t \mid s)$ as the next state of the Markov chain; otherwise, the Markov chain stays in state s.

Metropolis–Hastings algorithm

Step 0. Choose probability mass functions $\{q(t \mid s), t \in S\}$ for $s \in S$ such that the Markov matrix with the $q(t \mid s)$ as elements is irreducible. Choose a starting state s_0. Let $n := 1$.

Step 1. Generate a candidate state t_n from the probability mass function $\{q(t \mid s_{n-1}), t \in S\}$. Calculate the acceptance probability

$$\alpha = \min \left[\frac{\pi(t_n)q(s_{n-1} \mid t_n)}{\pi(s_{n-1})q(t_n \mid s_{n-1})}, 1 \right].$$

Step 2. Generate a random number u from $(0, 1)$. If $u \leq \alpha$, accept t_n and let $s_n := t_n$; otherwise, $s_n := s_{n-1}$.

Step 3. $n := n + 1$. Repeat step 1 with s_{n-1} replaced by s_n.

Note that when the chosen probability densities $q(t \mid s)$ are symmetric, that is, $q(t \mid s) = q(s \mid t)$ for all $s, t \in S$, then the acceptance probability α in Step 1 reduces to

$$\alpha = \min \left(\frac{\pi(t_n)}{\pi(s_{n-1})}, 1 \right).$$

In applications of the algorithm, one typically want to estimate $E[h(X)]$ for a given function $h(x)$, where X is a random variable with $\pi(s)$ as its probability distribution. If the states $s_1, s_2, \ldots, s_m$ are generated by the Metropolis–Hastings algorithm for a sufficiently large m, then $E[h(X)] = \sum_{s \in S} h(s)\pi(s)$ is estimated by

$$\frac{1}{m} \sum_{k=1}^{m} h(s_k).$$

This estimate is based on a law of large numbers result for Markov chains, saying that $\lim_{m \to \infty} \frac{1}{m} \sum_{k=1}^{m} h(s_k) = E[h(X)]$. A heuristic explanation is as follows. The probability $\pi(s)$ can be interpreted as the long-run fraction of transitions into state s and so, for m large, $\pi(s) \approx \frac{m(s)}{m}$, where $m(s)$ is the number of times that state s occurs among the sequence $s_1, s_2, \ldots, s_m$. This gives $\sum_{s \in S} h(s)\pi(s) \approx \frac{1}{m} \sum_{s \in S} h(s)m(s) = \frac{1}{m} \sum_{k=1}^{m} h(s_k)$.

The Metropolis–Hastings algorithm directly extends to the case of a probability density $\pi(s)$ on a (multi-dimensional) continuous set S, where you want to calculate $\int_{s \in S} h(s)\pi(s)\, ds$ for the case that the density $\pi(s)$ is only known up to a multiplicative constant.

What are the best options for the proposal functions $q(t\,|\,s)$? There are two general approaches: independent chain sampling and random walk chain sampling.

(a) In independent chain sampling the candidate state t is drawn independently of the current state s of the Markov chain, that is, $q(t \mid s) = g(t)$ for some proposal density $g(x)$.

(b) In random walk chain sampling the candidate state t is the current state s plus a draw from a random variable Z that does not depend on the current state. In this case, $q(t \mid s) = g(t - s)$ with $g(z)$ the density of the random variable Z. If $g(z) = g(-z)$ for all z, the proposal density is symmetric and, as noted before, the acceptance probability reduces to $\alpha(t \mid s) = \min(\pi(t)/\pi(s), 1)$.

It is very important to have a well-mixing Markov chain that explores the state space S adequately and sufficiently fast.

Gibbs sampler

The Gibbs sampler is a special case of Metropolis–Hastings method. This sampler is frequently used in Bayesian statistics. To introduce the method, consider the discrete random vector $(X_1, \ldots, X_d)$ with the multivariate distribution

$$\pi(x_1, \ldots, x_d) = P(X_1 = x_1, \ldots, X_d = x_d).$$

The univariate conditional distributions of $(X_1, \ldots, X_d)$ are

$$\pi_k(x \mid x_1, \ldots, x_{k-1}, x_{k+1}, \ldots, x_d)$$
$$= P(X_k = x \mid X_1 = x_1, \ldots, X_{k-1} = x_{k-1}, X_{k+1} = x_{k+1}, \ldots, X_d = x_d).$$

The key to the Gibbs sampler is the assumption that the univariate conditional distributions are fully known. The Gibbs sampler generates random draws from the univariate distributions and defines a Markov chain with $\pi(x_1, \ldots, x_d)$ as its unique equilibrium distribution. In each iteration, one component of the state vector is randomly chosen and adjusted. The algorithm is a special case of the Metropolis–Hastings method with

$$q(\mathbf{y} \mid \mathbf{x}) = \frac{1}{d} P\left(X_k = y \mid X_j = x_j \text{ for } j = 1, \ldots, d \text{ with } j \neq k\right),$$

for two states $\mathbf{x}$ and $\mathbf{y}$ such that $\mathbf{x} = (x_1, \ldots, x_{k-1}, x_k, x_{k+1}, \ldots, x_d)$ and $\mathbf{y} = (x_1, \ldots, x_{k-1}, y, x_{k+1}, \ldots, x_d)$. Since $q(\mathbf{x} \mid \mathbf{y}) = c\pi(\mathbf{x})$ and $q(\mathbf{y} \mid \mathbf{x}) = c\pi(\mathbf{y})$ with $c^{-1} = d\,P\left(X_k = y \mid X_j = x_j \text{ for } j \neq k\right)$, the reversibility property $q(\mathbf{x} \mid \mathbf{y})\pi(\mathbf{y}) = q(\mathbf{y} \mid \mathbf{x})\pi(\mathbf{x})$ is satisfied, and so the acceptance probability is 1. The Gibbs sampler is as follows:

Step 0. Choose a starting state $\mathbf{x} = (x_1, \ldots, x_d)$.
Step 1. Generate a random integer k from $\{1, \ldots, d\}$. Simulate a random draw y from $\pi_k(x \mid x_1, \ldots, x_{k-1}, x_{k+1}, \ldots, x_d)$. Define state $\mathbf{y}$ by $\mathbf{y} = (x_1, \ldots, x_{k-1}, y, x_{k+1}, \ldots, x_d)$.
Step 2. The new state $\mathbf{x} := \mathbf{y}$. Return to step 1 with $\mathbf{x}$.

In practice one usually uses a slightly modified Gibbs sampler, where in each iteration all components of the state vector are successively adjusted rather than a single component.

Solutions to Selected Problems

Fully worked-out solutions to a number of problems are given. Including worked-out solutions is helpful for students who use the book for self-study and stimulates active learning. Make sure you try the problems before looking to the solutions.

2.4. Two methods will be presented to solve this problem by calculating the complementary probability that there is no ace among the four cards. It is always helpful if you can check the solution using alternative solution methods.

Solution method 1: This method uses an unordered sample space. The sample space consists of all possible combinations of four different cards. This sample space has $\binom{52}{4}$ equally likely elements. The number of elements for which there is no ace among the four cards is $\binom{48}{4}$. Thus the probability of getting no ace is $\binom{48}{4}/\binom{52}{4} = 0.7187$ and so the desired probability is $1 - 0.7187 = 0.2813$.

Solution method 2: This method uses an ordered sample space. The sample space consists of all possible orderings of the 52 cards. The sample space has $52!$ equally likely elements. The number of possible orderings for which there is no ace among the first four cards in the ordering is $48 \times 47 \times 46 \times 45 \times 48!$. The probability of getting no ace can also be found as $(48 \times 47 \times 46 \times 45 \times 48!)/52! = 0.7187$.

2.5. It is often helpful to rephrase a probability problem in another context. The probability that the other card has no free drinks either is nothing else than the probability of getting 10 heads and 10 tails in 20

159

coin tosses (think about it!). To calculate the latter probability, use as sample space the set consisting of all possible sequences of H's and T's of length 20. The sample space has 2^{20} equally likely elements. There are $\binom{20}{10}$ sequences having 10 H's and 10 T's. Thus the probability of getting 10 heads and 10 tails is $\binom{20}{10}/2^{20} = 0.1762$. Therefore the sought-after probability is 0.1762.

2.6. The strategy is that the daughter (father) opens door 1 (2) first. If the key (car) is behind door 1 (2), the daughter (father) goes on to open door 2 (1). If the goat is after the first opened door, door 3 is opened as second. Then four of the six possible configurations of the car, key and goat are favorable: the configurations (car, key, goat), (car, goat, key), (key, car, goat) and (goat, key, car) are winning, and the two configurations (key, goat, car) and (goat, car, key) are losing.

2.7. If the sets A and B are not disjoint, then $P(A) + P(B)$ counts twice the probability of the set of outcomes that belong to both A and B. Therefore $P(A \text{ and } B)$ should be subtracted from $P(A) + P(B)$. Let A be the event that the chosen card is a heart and B be the event that it is an ace, then $P(A) = \frac{13}{52}$, $P(B) = \frac{4}{52}$ and $P(A \text{ and } B) = \frac{1}{52}$, and so $P(A \text{ or } B) = \frac{13}{52} + \frac{4}{52} - \frac{1}{52} = \frac{16}{52}$.

2.8. The probability is $\sum_{k=1}^{\infty} (1 - p_1 - p_2)^{k-1} p_1 = \frac{p_1}{p_1+p_2}$.

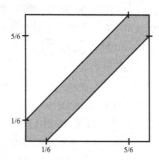

2.9. Translate the problem into choosing a point at random inside the unit square. The probability that the two persons will meet within 10 minutes of each other is equal to the probability that a point chosen at random in the unit square will fall inside the shaded region, see the figure. The area of the shaded region is calculated as $1 - \frac{5}{6} \times \frac{5}{6} = \frac{11}{36}$.

Dividing this area by the area of the unit square, you get that the desired probability is $\frac{11}{36}$.

2.11. Imagine that the two cards are picked one by one. Then, using the product rule, the probability of getting two red cards is $\frac{14}{21} \times \frac{13}{20} = \frac{13}{30}$ and the probability of getting one red card and one black card is $\frac{14}{21} \times \frac{7}{20} + \frac{7}{21} \times \frac{14}{20} = \frac{14}{30}$. Alternatively, the solution can be obtained by using the urn model from Section 1.1: the probability of getting two red cards is $\binom{14}{2}/\binom{21}{2} = \frac{13}{30}$ and the probability of getting one red card and one black card is $\binom{14}{1}\binom{7}{1}/\binom{21}{2} = \frac{14}{30}$.

2.20. Let A be the event that there is originally a piranha in the bowl and B be the event that the removed fish is a piranha. The sought-after probability is $P(A \mid B)$. This probability can be calculated from the basic formula $P(A \mid B) = \frac{P(A)P(B \mid A)}{P(B)}$. The values of $P(A)$ and $P(B \mid A)$ are $\frac{1}{2}$ and 1. Letting $\overline{A}$ be the event that the original fish in the bowl is not a piranha. Then, by the law of conditional probability,

$$P(B) = P(B \mid A)P(A) + P(B \mid \overline{A})P(\overline{A}) = 1 \times \frac{1}{2} + \frac{1}{2} \times \frac{1}{2} = \frac{3}{4}.$$

Therefore $P(A \mid B) = \frac{1/2}{3/4} = \frac{2}{3}$. An alternative way to get this answer is to use Bayes' rule in odds form from Section 2.4.

2.21. Let A be the event that you ever win the jackpot by buying a ticket only once. The event A can only occur if one of the disjoints events B_1 and B_2 occurs, where B_1 is the event of having six correctly predicted numbers on your first ticket and B_2 is the event of having exactly two correctly predicted numbers on your first ticket. Then, $P(A) = P(A \mid B_1)P(B_1) + P(A \mid B_2)P(B_2)$. Using the urn model from Section 1.1, $P(B_1) = 1/\binom{45}{6}$ and $P(B_2) = \binom{6}{2}\binom{39}{4}/\binom{45}{6}$. Also, $P(A \mid B_1) = 1$ and $P(A \mid B_2) = P(A)$. Thus $P(A) = 1/\binom{45}{6} + P(A) \times \binom{6}{2}\binom{39}{4}/\binom{45}{6}$, and so $P(A) = 1.447 \times 10^{-7}$.

2.26. Let the hypothesis H be the event that the standard die was picked and let the evidence E_1 be the event that the first roll of the picked die has the outcome 6. The prior probabilities are $P(H) =$

$P(\overline{H}) = \frac{1}{2}$ and the likelihood ratio has $P(E_1 \mid H) = \frac{1}{6}$ and $P(E_1 \mid \overline{H}) = \frac{1}{3}$. Thus the posterior odds of the hypothesis H are

$$\frac{P(H \mid E_1)}{P(\overline{H} \mid E_1)} = \frac{1/2}{1/2} \times \frac{1/6}{1/3} = \frac{1}{2},$$

which gives the updated value $\frac{1/2}{1+1/2} = \frac{1}{3}$ for the probability that the standard die was picked. The second question can be answered in two ways. After the first roll has been done but before the second roll will be done, you take the posterior probabilities $P(H \mid E) = \frac{1}{3}$ and $P(\overline{H} \mid E) = \frac{2}{3}$ as the prior probabilities for $P(H)$ and $P(\overline{H})$. Doing so and letting E_2 be the event that the second roll has outcome 6, you get

$$\frac{P(H \mid E_2)}{P(\overline{H} \mid E_2)} = \frac{1/3}{2/3} \times \frac{1/6}{1/3} = \frac{1}{4},$$

and so the newly updated value of the probability that the standard die was picked is $\frac{1}{5}$. Alternatively, this probability can be calculated by letting the evidence $E_{1,2}$ be the event that each of the first two rolls of the picked die has outcome 6. Then, before the first two rolls are done, the priors $P(H)$ and $P(\overline{H})$ are $\frac{1}{2}$ and the likelihood ratio has $P(E_{1,2} \mid H) = \frac{1}{6} \times \frac{1}{6}$ and $P(E_{1,2} \mid \overline{H}) = \frac{1}{3} \times \frac{1}{3}$. This leads to

$$\frac{P(H \mid E_{1,2})}{P(\overline{H} \mid E_{1,2})} = \frac{1/2}{1/2} \times \frac{1/36}{1/9} = \frac{1}{4},$$

which gives again the update $\frac{1}{5}$ for the probability that the standard die was picked. The Bayesian approach has the feature that you can continuously update your beliefs as information accrues. Verify yourselves that the updated value of the probability that the standard die was picked becomes $\frac{1}{9}$ after a third roll with outcome 2. *Note:* this problem nicely illustrates the Bayesian view that probabilities represent the knowledge an observer has about the state of nature of a physical object.

2.27. Let the hypothesis H be the event that the person has the disease and the evidence E be the event that he has tested positive. The prior probabilities are $P(H) = 0.001$ and $P(\overline{H}) = 0.999$. Also, the likelihood ratio has $P(E \mid H) = 0.99$ and $P(E \mid \overline{H}) = 0.01$. Thus the posterior odds of hypothesis H are $\frac{0.001}{0.999} \times \frac{0.99}{0.01} = \frac{11}{111}$, and so the posterior probability of hypothesis H is $\frac{11/111}{1+11/111} = 0.0902$. In other words, the probability that the person has the disease is only 9.02% if the first test is positive. The low value of this probability may be surprising when not taking into account the *base rate*: most positive tests come from people who don't have the disease. If a second independent test also gives a positive test result, then use the posterior probabilities 0.0902 and $1 - 0.0902 = 0.9098$ as new prior probabilities for $P(H)$ and $P(\overline{H})$. This leads to the new posterior odds $\frac{0.0902}{0.9098} \times \frac{0.99}{0.01} = 9.815$. Thus, after a second positive test, the probability that the person has the disease is $\frac{9.815}{1+9.815} \times 100 = 90.75\%$.

The posterior probabilities can also be heuristically argued be the expected frequency approach. This is first done for the case that 0.1% of the test population has the disease. Think of a large number of people, say 10 000. Of these 10 000 people, 10 will have the disease on average, and 9 990 will not. Of the 10 persons with the disease $0.99 \times 10 = 9.9$ will test positive on average, and of the 9 990 persons with no disease $0.01 \times 9\,990 = 99.9$ will test positive on average. Thus the posterior probability of the disease is equal to $9.9/(9.9 + 99.9) = 0.0902$ when the first test is positive. Similarly, if a second test is also positive, you get the posterior probability $892.98/(892.98+90.98)=0.9075$.

2.29. Let the random variable X be the value of the stock two weeks later. The underlying sample space consists of the four equally likely outcomes II, ID, DI and DD, where I stands for increase and D for decrease. The random variable X takes on the respective values $1.05 \times 1.05 \times 100 = 110.25$, $1.05 \times 0.96 \times 100 = 100.80$, $0.96 \times 1.05 \times 100 = 100.80$, and $0.96 \times 0.96 \times 100 = 92.16$ for these four outcomes. Thus $P(X = 110.25) = 0.25$, $P(X = 100.80) = 0.50$, and $P(X = 92.16) = 0.25$.

2.37. Let X be the number of cards that need to be turned over. There are several ways to calculate $E(X)$. The straightforward way is $E(X) = \sum_{k=0}^{49} k \left[\binom{48}{k-1}/\binom{52}{k-1} \right] \times \frac{4}{52-(k-1)} = 10.6$, using the result below Example 2.7. An alternative way is to write $X = 1 + \sum_{j=1}^{48} I_j$, where $I_j = 1$ if the jth non-ace card in the deck appears before the first ace and $I_j = 0$ otherwise. For each j, you have $P(I_j = 1) = \frac{1 \times 4!}{5!} = \frac{1}{5}$, as can be seen by considering the positions of the jth non-ace card and the four aces relative to each other. Using the linearity of expectation, you then get $E(X) = 1 + 48 \times \frac{1}{5} = 10.6$. The alternative derivation can also be used to show that $E(X_i) = i + 48\left(\frac{i \times 4!}{5!}\right) = i \times 10.6$ for $i = 1, \ldots, 4$, where X_i is the number of cards that need to be turned over until an ace appears for the ith time.

2.38. $E(X) = m\binom{10}{m}/\binom{12}{m}$ is maximal for $m = 4$ with $E(X) = \frac{56}{33}$.

2.42. This problem is an application of the balls-and-bins model with $n = \binom{42}{6}$ bins (six-number combinations) and $b = 5\,000\,000$ balls (tickets). By the same reasoning as in Example 2.13, you find that the expected value of the number of empty bins is $n \times \left(\frac{n-1}{n}\right)^b$, which equals $2\,022\,388$ for $n = 5\,245\,786$ bins and $b = 5\,000\,000$ balls. Thus the expected number of *different* six-number combinations filled in is $5\,245\,786 - 2\,022\,388 = 3\,223\,398$ (there are surprisingly many duplicates among the tickets filled in; the probability of the jackpot falling is only $3\,223\,398/5\,245\,786 = 0.6145$). *Note:* the insightful and accurate approximation $n \times \left(\frac{n-1}{n}\right)^b \approx ne^{-b/n}$ applies for n large.

2.43. Let X be the largest number drawn. Then $P(X \leq k) = \binom{k}{6}/\binom{49}{6}$. This gives $E(X) = \sum_{k=6}^{49} k\,[P(X \leq k) - P(X \leq k-1)] = 42.857$. This problem is related to the German Tank Problem in which the unknown total number of tanks had to estimated from a few tank numbers.

2.44. By the definitions of expected value and variance, $E(X) = 0 \times (1-p) + 1 \times p = p$ and $\sigma^2(X) = (0-p)^2 \times (1-p) + (1-p)^2 \times p = p(1-p)$ and so $\sigma(X) = \sqrt{p(1-p)}$.

2.46. Let S be the number of boxes needed to get a complete set of cards. Then S can be written as $S = Y_0 + Y_1 + \cdots + Y_{49}$, where Y_i is the number of purchases needed to go from i distinct cards to $i+1$ distinct cards. Then Y_i is geometrically distributed with parameter $p_i = \frac{50-i}{50}$ and so $E(Y_i) = \frac{50}{50-i}$. This gives $E(S) = 50 \sum_{k=1}^{50} \frac{1}{k} = 224.96$.

2.48. Suppose the strategy is to stop as soon as you have picked a number larger than or equal to r (≤ 25). The number of trials needed is geometrically distributed with success probability $p = \frac{25-r+1}{25}$ (and expected value $\frac{25}{25-r+1}$). Each of the values $r, r+1, \ldots, 25$ is equally likely for your payout. Thus the expected net payoff is $\sum_{k=r}^{25} k \times \frac{1}{25-r+1} - \frac{25}{25-r+1} \times 1$, which can be simplified as $\frac{1}{2}(25+r) - \frac{25}{25-r+1}$. This expression has the maximal value 18.4286 for $r = 19$.

2.49. Let the number a be your guess and Y be the randomly chosen number. Your expected winning is $E[g(Y)]$, where $g(y) = a^2$ for $y > a$ and $g(y) = 0$ otherwise. Then $E[g(Y)] = \sum_{k=a+1}^{100} a^2 P(Y = k)$. Since $P(Y = k) = 1/100$ for all k, you get $E[g(Y)] = (100-a)a^2/100$. This expression is maximal for $a = 67$ with 1481.37 as maximum value.

2.53. Let the random variable S be the number of rolls needed to get all six possible outcomes. Then $S = \sum_{i=0}^{5} X_i$, where X_i is geometrically distributed with parameter $\frac{6-i}{6}$ and the X_i are independent. Using the results of Problem 2.45, $\sigma^2(S) = \sum_{i=0}^{5} \left(\frac{6}{6-i}\right)^2 (1 - \frac{6-i}{6}) = 38.99$ and so $\sigma(S) = 6.244$.

3.2. The probability is $\sum_{k=3}^{5} \binom{5}{k} \left(\frac{4}{25}\right)^k \left(\frac{21}{25}\right)^{5-k} = 0.0318$.

3.7. We only verify the second answer, using Bonferroni's inequality $\sum_{i=1}^{n} P(A_i) - \sum_{i=1}^{n-1} \sum_{j=i+1}^{n} P(A_i \text{ and } A_j) \le P(A_1 \text{ or } \cdots \text{ or } A_n) \le \sum_{i=1}^{n} P(A_i)$. For any i, let A_i be the event that the outcome i appears at least three times in the roll of the six dice. Then, $P(A_i)$ equals $\sum_{k=3}^{6} \binom{6}{k} \left(\frac{1}{6}\right)^k \left(\frac{5}{6}\right)^{6-k} = 0.3737$. Also, $P(A_i \text{ and } A_j) = \binom{6}{3} \left(\frac{1}{6}\right)^3 \left(\frac{1}{6}\right)^3$ for $i \ne j$. This gives $0.3737 - 0.0064 \le Q \le 0.3737$ for the desired probability $Q = P(A_1 \text{ or } \cdots \text{ or } A_6)$.

3.9. Let the random variable G be your end capital and X be the number of times that heads shows up in the 10 bets. Then $G = G(X)$, where the function $g(x) = 1.7^x \times 0.5^{10-x} \times 100$ for $x = 0, 1, \ldots 10$. The probability that G takes on the value $g(x)$ is $P(X = x) = \binom{10}{x} 0.5^x 0.5^{10-x}$. The function $g(x)$ is increasing and the first value of x such that $g(x) > 100$ is $x = 6$. Thus $P(G > 100) = P(X \ge 6) = 0.3770$. The explanation that this probability is so small is that $1.7 \times 0.5 < 1$ (in the most likely scenario of 5 heads and 5 tails, your bankroll decreases from \$100 to $0.85^5 \times 100 = \$44.37$.) The lesson is: do not simply rely on averages in situations of risk, but use probability distributions!

3.16. An appropriate model is the Poisson model. The sought probability can be estimated as $e^{-20/400} \approx 0.951$.

3.20. The length X of a gestation period is $N(\mu, \sigma^2)$ distributed with $\mu = 280$ days and $\sigma = 10$ days. The probability that a birth is more than 15 days overdue is $1 - P(X \leq 295)$. This probability can be evaluated as $1 - P\left(\frac{X-280}{10} \leq \frac{295-280}{10}\right) = 1 - \Phi(1.5) = 0.0668$, using the basic result that $\frac{X-\mu}{\sigma}$ is $N(0,1)$ distributed. In other words, the proportion of births that are more than 15 days overdue is 6.68%.

3.23. Let X denote the demand for the item. The normally distributed random variable X has an expected value of $\mu = 100$ and satisfies $P(X > 125) = 0.05$. To find the unknown standard deviation σ of X, write $P(X > 125) = 0.05$ as $P\left(\frac{X-100}{\sigma} > \frac{125-100}{\sigma}\right) = 1 - \Phi\left(\frac{25}{\sigma}\right) = 0.05$. Thus $\Phi\left(\frac{25}{\sigma}\right) = 0.95$. The percentile $\xi_{0.95} = 1.645$ is the unique solution to the equation $\Phi(x) = 0.95$. Thus $\frac{25}{\sigma} = 1.645$, which gives $\sigma = 15.2$.

3.28. Letting $X_1, \ldots, X_n$ be independent random variables with $P(X_i = 1) = P(X_i = -1) = 0.5$ for all i, the random variable D_n can be represented as $D_n = |X_1 + \cdots + X_n|$. Since $E(X_i) = 0$ and $\sigma(X_i) = 1$ for all i, the expected value and standard deviation of $X_1 + \cdots + X_n$ are 0 and $\sqrt{n}$. By the central limit theorem, the random variable $X_1 + \cdots + X_n$ is approximately $N(0, n)$ distributed for n large. Let's now calculate $E(|V|)$ if V is $N(0, \sigma^2)$ distributed:

$$E(|V|) = \frac{1}{\sigma\sqrt{2\pi}} \int_{-\infty}^{\infty} |v| \, e^{-\frac{1}{2}v^2/\sigma^2} \, dv = \frac{2}{\sigma\sqrt{2\pi}} \int_0^{\infty} v \, e^{-\frac{1}{2}v^2/\sigma^2} \, dv.$$

By the change of variable $w = (v/\sigma)^2$, the latter integral for $E(|V|)$ becomes $E(|V|) = \frac{\sigma}{\sqrt{2\pi}} \int_0^{\infty} e^{-\frac{1}{2}w} \, dw = \frac{2\sigma}{\sqrt{2\pi}}$, yielding the desired result for $E(D_n)$.

3.30. Let the random variable X_i be the dollar amount the casino loses on the ith bet for $i = 1, \ldots, 2\,500$. The X_i are independent random variables with $P(X_i = 10) = \frac{18}{37}$ and $P(X_i = -5) = \frac{19}{37}$. Then $E(X_i) = \frac{85}{37}$ and $\sigma(X_i) = \frac{45}{37}\sqrt{38}$. The total dollar amount lost by the casino is $\sum_{i=1}^{2\,500} X_i$. By the central limit theorem, this sum is approximately $N(\mu, \sigma^2)$ distributed with $\mu = 2\,500 \times \frac{85}{37}$ and $\sigma = 50 \times \frac{45}{37}\sqrt{38}$. The

casino will lose no more than 6500 dollars with a probability of about
$\Phi\left(\frac{6\,500-\mu}{\sigma}\right) = 0.978$.

3.33. By the memoryless property of the exponential distribution, your
waiting time at the bus stop is more than s minutes with probability
$e^{-\frac{1}{10}s}$. Solving $e^{-\frac{1}{10}s} = 0.05$ gives $s = 29.96$. Thus you should leave
home about 35 minutes before 7:45 a.m.

3.34. The number of births between twelve o'clock midnight and six
o'clock in the morning has a Poisson distribution with an expected value
of $6 \times \frac{5}{24} = \frac{5}{4}$. Thus the probability of having more than two births is
$1 - \sum_{k=0}^{2} e^{-5/4}\frac{(5/4)^k}{k!} = 0.1315$.

3.35. Your win probability is the probability of having exactly one signal
in (s, T). This probability is $e^{-\lambda(T-s)}\lambda(T-s)$, using the fact that the
number of signals in any time interval of length t is Poisson distributed
with expected value λt. Putting the derivative of $e^{-\lambda(T-s)}\lambda(T-s)$ with
respect to s equal to zero, you get that the optimal value of s is $T - \frac{1}{\lambda}$.
The maximal win probability is e^{-1}.

3.36. Let the random variable X_i be the score on exam i for $i = 1, 2$.
(a) The density of X_1 is the $N(\mu_1, \sigma_1^2) = N(75, 12)$ density. This gives
$P(X_1 \geq 80) = 1 - \Phi\left(\frac{80-75}{12}\right) = 0.3385$.
(b) The random variable $X_1 + X_2$ is $N(\mu_1 + \mu_2, \sigma_1^2 + \sigma_2^2 + 2\rho\sigma_1\sigma_2) =$
$N(140, 621)$ distributed. Thus $P(X_1 + X_2 > 150) = 1 - \Phi\left(\frac{150-140}{\sqrt{621}}\right) =$
0.3441.
(c) The random variable $X_2 - X_1$ is $N(\mu_2 - \mu_1, \sigma_1^2 + \sigma_2^2 - 2\rho\sigma_1\sigma_2) =$
$N(-10, 117)$ distributed. Thus $P(X_2 > X_1) = 1 - \Phi\left(\frac{10}{\sqrt{117}}\right) = 0.1776$.
(d) The conditional distribution of X_2 given that $X_1 = 80$ equals the
$N(65 + 0.7 \times \frac{15}{12} \times (80 - 75), 225 \times (1 - 0.49)) = N(69.375, 114.75)$
distribution. Thus $P(X_2 > X_1 | X_1 = 80) = 1 - \Phi\left(\frac{80-69.375}{\sqrt{114.75}}\right) = 0.1606$.

4.3. Let A be the event that you get back your own exam and B be
the event that none of the other 14 students gets back their own exams.
Then, the desired probability is $P(A \text{ and } B)$. The product rule gives
$P(A \text{ and } B) = P(A)P(A \mid B)$. The probability $P(A)$ is $\frac{1}{15}$, while the
conditional probability $P(A \mid B)$ is nothing else than the probability that

no child picks his/her own present in the Santa Claus problem with 14 children. Thus $P(A \mid B) = e^{-1}$, and so the probability that you are the only person who gets back the own exam is $\frac{e^{-1}}{15} = 0.0245$.

6.1. The probability of getting a fine is equal to $p_{22}^{(4)} + p_{23}^{(4)} = 0.625$.

6.2. Use a Markov model with four states SS, SR, RS and RR describing the weather of yesterday and today and with one-step transition probabilities

$$
\begin{array}{c|cccc}
\text{from}\backslash\text{to} & SS & SR & RS & RR \\
\hline
SS & 0.9 & 0.1 & 0 & 0 \\
SR & 0 & 0 & 0.5 & 0.5 \\
RS & 0.7 & 0.3 & 0 & 0 \\
RR & 0 & 0 & 0.45 & 0.55
\end{array}
$$

The probability of having sunny weather five days from now if it rained both today and yesterday is $p_{RR,SS}^{(5)} + p_{RR,RS}^{(5)}$. Calculating

$$
\mathbf{P}^5 = \begin{pmatrix}
0.7178 & 0.0977 & 0.0879 & 0.0966 \\
0.6151 & 0.1024 & 0.1325 & 0.1501 \\
0.6837 & 0.0997 & 0.1024 & 0.1142 \\
0.6089 & 0.1028 & 0.1351 & 0.1532
\end{pmatrix},
$$

you find that the desired probability is $0.6089 + 0.1351 = 0.7440$.

6.3. This problem is an instance of the balls-and-bins model, as can be seen by imagining that the passengers inform the bus driver one by one of their destination before boarding the bus. The state is the current number of known stops. The p_{ij} are $p_{01} = 1$, $p_{i,i} = \frac{i}{7}$ and $p_{i,i+1} = 1 - \frac{i}{7}$ for $1 \leq i \leq 6$, $p_{77} = 1$, and $p_{ij} = 0$ otherwise. Calculating $\mathbf{P}^{25}$ gives $(p_{0j}^{(25)}) = (0, 0.0000, 0.0000, 0.0000, 0.0046, 0.1392, 0.8562)$.

6.4. Let the indicator variable I_k be 1 if it is sunny k days from now and be 0 otherwise, given that it is cloudy today. Then $P(I_k = 1) = p_{CS}^{(k)}$. Since $E(I_k) = 1 \times P(I_k = 1)$, you have $E\left(\sum_{k=1}^{7} I_k\right) = \sum_{k=1}^{7} E(I_k) = \sum_{k=1}^{7} p_{CS}^{(k)}$. This leads to $E\left(\sum_{k=1}^{7} I_k\right) = 4.049$.

6.6. Use a Markov chain with state space $I = \{0, 1, \ldots, 6\}$, where state $i \neq 0$ means that the last i rolls resulted in the first i digits of 123456 and state 0 is an auxiliary state. State 6 is taken as an absorbing state. The Markov chain has the one-step transition probabilities $p_{01} = \frac{1}{6}$, $p_{00} = \frac{5}{6}$, $p_{i,i+1} = p_{i1} = \frac{1}{6}$ and $p_{i0} = \frac{4}{6}$ for $i = 1, \ldots, 5$, $p_{66} = 1$, and the other $p_{ij} = 0$. Calculating $\mathbf{P}^{500}$ leads to $p_{06}^{(500)} = 0.0106$.

6.11. Let state i mean that the dragon has i heads for $i = 0, 1, \ldots, 4$ and state 5 mean that the dragon has 5 or 6 heads. The states 0 and 5 are taken as absorbing states. The one-step transition probabilities are $p_{00} = 1$, $p_{i,i-1} = 0.5$, $p_{i,i+1} = 0.35$ and $p_{i,i+2} = 0.15$ for $1 \leq i \leq 3$, $p_{43} = p_{45} = 0.5$, $p_{55} = 1$ and the other $p_{ij} = 0$. For any starting state, the process will ultimately absorbed in either state 0 or state 5. The absorption probabilities can be obtained by calculating $\mathbf{P}^n$ for n sufficiently large. Trying several values of n, it was found that $n = 50$ is large enough to have convergence of all $p_{ij}^{(n)}$ in four or more decimals:

$$
\mathbf{P}^{50} = \begin{pmatrix}
1 & 0 & 0 & 0 & 0 & 0 \\
0.7112 & 0.0000 & 0.0000 & 0.0000 & 0.0000 & 0.2888 \\
0.4790 & 0.0000 & 0.0000 & 0.0000 & 0.0000 & 0.5210 \\
0.2903 & 0.0000 & 0.0000 & 0.0000 & 0.0000 & 0.7097 \\
0.1451 & 0.0000 & 0.0000 & 0.0000 & 0.0000 & 0.8549 \\
0 & 0 & 0 & 0 & 0 & 1
\end{pmatrix}.
$$

Since $p_{20}^{(50)} = 0.4790$, your probability of winning is 47.9%. Alternatively, this probability can be calculated by solving four linear equations. To do so, define f_i as the probability of ever getting absorbed in state 0 when the starting state is i. By definition, $f_0 = 1$ and $f_5 = 0$. By conditioning on the next state after state i and using the law of conditional probability, you get the four linear equations $f_i = 0.5f_{i-1} + 0.35f_{i+1} + 0.15f_{i+2}$ for $i = 1, 2, 3$ and $f_4 = 0.5f_3 + 0.5f_5$. The solution is $f_1 = 0.7112$, $f_2 = 0.4790$, $f_3 = 0.2903$ and $f_4 = 0.1451$.

6.14. This problem is a gem for teaching absorbing Markov chains. The state of the Markov chain is described by the triple (i, r_1, r_2), where i denotes the number of smashed eggs, r_1 is the number of raw eggs picked by the guest and r_2 is the number of raw eggs picked by the host of the game. The states satisfy $0 \leq i \leq 11$ and $r_1 + r_1 \leq 3$. The process starts in state $(0, 0, 0)$ and ends when one of the absorbing states $(i, 2, 0)$, $(i, 2, 1)$, $(i, 0, 2)$, or $(i, 1, 2)$ is reached. The guest loses the game if the game ends in a state $(i, 2, 0)$ or $(i, 2, 1)$ with i odd. In a non-absorbing state (i, r_1, r_2) with i even, the guest picks an egg and the process goes either to state $(i + 1, r_1 + 1, r_2)$ with probability $\frac{4 - r_1 - r_2}{12 - i}$ or to state $(i + 1, r_1, r_2)$ with probability $1 - \frac{4 - r_1 - r_2}{12 - i}$. In a non-absorbing state (i, r_1, r_2) with i odd, the host picks an egg and the process goes either to state $(i + 1, r_1, r_2 + 1)$ with probability $\frac{4 - r_1 - r_2}{12 - i}$ or to state $(i + 1, r_1, r_2)$ with probability $1 - \frac{4 - r_1 - r_2}{12 - i}$. This sets the matrix **P** of one-step transition probabilities. The probability that the guest will lose can be computed by calculating $\mathbf{P}^{11}$. This requires that the states are ordered in a one-dimensional array. It is easier to use a recursion to calculate the probability of the guest losing the game. To that end, you reason in the same way as in the solution of Problem 6.11. For any state (i, r_1, r_2), let $p(i, r_1, r_2)$ be the probability that the guest will lose if the process starts in state (i, r_1, r_2). The goal is to find $p(0, 0, 0)$. This probability can be calculated by a recursion with the boundary conditions $p(i, 2, 0) = p(i, 2, 1) = 1$ and $p(i + 1, 0, 2) = p(i + 1, 1, 2) = 0$ for $i = 3$, 5, 7, 9 and 11. The recursion is

$$p(i, r_1, r_2) = \frac{4 - r_1 - r_2}{12 - i} p(i+1, r_1+1, r_2) + \left(1 - \frac{4 - r_1 - r_2}{12 - i}\right) p(i+1, r_1, r_2)$$

for $i = 0$, 2, 4, 6, 8 and 10, and

$$p(i, r_1, r_2) = \frac{4 - r_1 - r_2}{12 - i} p(i+1, r_1, r_2+1) + \left(1 - \frac{4 - r_1 - r_2}{12 - i}\right) p(i+1, r_1, r_2)$$

for $i = 1$, 3, 5, 7, 9 and 11. The recursive computations lead to the value $\frac{5}{9}$ for the probability that the guest of the show will lose the game. It is interesting to note that the game is fair for the case of three raw eggs and nine boiled eggs.

6.17. The first thought might be to use a Markov chain with 16 states. However, a Markov chain with two states 0 and 1 suffices, where state 0 means that Linda and Bob are in different venues and state 1 means that they are in the same venue. The one-step-transition probability p_{01} is equal to $p_{01} = 2 \times 0.4 \times \left(0.6 \times \frac{1}{3}\right) + \left(0.6 \times \frac{2}{3}\right) \times \left(0.6 \times \frac{1}{3}\right) = 0.24$, where the first term refers to the probability that exactly one of the two persons does not change of venue and the other person goes to the venue of that person, and the second term refers to the probability that both persons change of venue and go the same venue. Similarly, $p_{11} = 0.4 \times 0.4 + 0.6 \times \left(0.6 \times \frac{1}{3}\right) = 0.28$. Further, $p_{00} = 1 - p_{01} = 0.76$ and $p_{10} = 1 - p_{11} = 0.72$. Solving the equations $\pi_0 = 0.76\pi_0 + 0.72\pi_1$ and $\pi_0 + \pi_1 = 1$ gives $\pi_0 = \frac{3}{4}$ and $\pi_1 = \frac{1}{4}$. The long-run fraction of weekends that Linda and Bob visit a same venue is $\pi_1 = \frac{1}{4}$.

6.18. (a) The solution of the balance equations $\pi_{SS} = 0.9\pi_{SS} + 0.7\pi_{RS}$, $\pi_{SR} = 0.1\pi_{SS} + 0.3\pi_{RS}$, $\pi_{RS} = 0.5\pi_{SR} + 0.45\pi_{RR}$ with the normalization equation $\pi_{SS} + \pi_{SR} + \pi_{RS} + \pi_{RR} = 1$ is given by $\pi_{SS} = 0.6923$, $\pi_{SR} = \pi_{RS} = 0.0989$, $\pi_{RR} = 0.1099$. The long-run fraction of sunny days is $\pi_{SS} + \pi_{RS} = 0.7912$.
(b) The long-average sales per day is $1\,000 \times 0.7912 + 500 \times 0.2088 = 895.60$ dollars. The standard deviations σ_1 and σ_2 are irrelevant for the long-run average sales.

Index

Printed in the United States
by Baker & Taylor Publisher Services